Animal Intelligence

Insights into the Animal Mind

Edited by R. J. Hoage and Larry Goldman

Smithsonian Institution Press
Washington, D.C., and London

The paper in this book meets the guide-
lines for permanence and durability of the
Committee on Production Guidelines for
Book Longevity of the Council on Library
Resources.

Library of Congress Cataloging in
Publication Data
Animal intelligence.
(National Zoological Park symposia for the
public series)

Papers were presented at a symposium held
at the National Zoological Park, Washington,
D.C., on April 9–10, 1983.
1. Animal intelligence—Congresses. I. Series.
QL785.A72443 1986 591.51 86-6708
ISBN 0-87474-541-1

National Zoological Park Symposia for the
Public

This series brings before the public a variety
of intriguing and controversial issues in wild-
life biology and conservation. Complicated
topics are addressed. Problems are analyzed.
In many cases, solutions are provided and
discussed. The goal of these books is to provide information to stimulate public awareness and discussion of the problems that
animals and their habitats face in a world increasingly dominated by human beings.

Cover drawing by Richard Swartz, modified
by the Office of Graphics and Exhibits, National Zoological Park.

To our children:
Chris, Theresa, Neal,
Kevin, Elliot, and David

Contents

Contributors

Marian Breland Bailey
Animal Behavior Enterprises, Inc.
4800 Albert Pike
Hot Springs, Arkansas 71913

Benjamin B. Beck
National Zoological Park
Smithsonian Institution
Washington, D.C. 20008

Colin G. Beer
Institute of Animal Behavior
Rutgers University
101 Warren Street
Newark, New Jersey 07102

David Challinor
Assistant Secretary for Science
Smithsonian Institution
Washington, D.C. 20560

Bennett G. Galef, Jr.
Department of Psychology
McMaster University
Hamilton, Ontario, Canada L8S 4K1

Carl Gans
Department of Zoology
University of Michigan
Ann Arbor, Michigan 48103

Larry Goldman
Department of Psychology
University of the District of
  Columbia
Washington, D.C. 20008

James L. Gould
Carol G. Gould
Department of Biology
Princeton University
Princeton, N.J. 08544

Robert J. Hoage
National Zoological Park
Smithsonian Institution
Washington, D.C. 20008

William Hodos
Department of Psychology
University of Maryland
College Park, Maryland 20742

Carolyn A. Ristau
The Rockefeller University
1230 York Avenue
New York, New York 10021

Duane M. Rumbaugh
Department of Psychology
Georgia State University
Atlanta, Georgia 30322

Sue Savage-Rumbaugh
Language Research Center
Yerkes Regional Primate Research
    Center
Emory University
Atlanta, Georgia 30322

Roger K. Thomas
Department of Psychology
University of Georgia
Athens, Georgia 30602

Stephen J. Vicchio
Department of Philosophy
Notre Dame College of Maryland
4701 North Charles Street
Baltimore, Maryland 21210

Preface

R. J. Hoage and Larry Goldman

From the earliest days of childhood we hear stories about animals who can talk, have feelings, and generally behave like humans. A quick review of children's literature, comic books, television programs, and films would show that our culture encourages the depiction of animals as thinking and acting intelligently. No wonder our young children are confused at times and insist on talking to animals they encounter—their pets, animals they see in the woods, in zoos, or on farms, not to mention their teddy bears and other toy animals. As we grow up we continue to relate to the animals we encounter as if they were intelligent. Those of us who live with a cat or dog or almost any kind of domesticated animal tend to be especially convinced of their intelligence. But how realistic are these evaluations of our pets? And what about wild animals and the many different animals we see at the zoo—how much do they understand? Can they learn from their individual experiences or is their behavior primarily instinctive, a direct result of their genetic inheritance? Scientists who study animal behavior are very interested in these kinds of questions. Differences in the learning abilities of various animals have long been a subject of inquiry, but recently there has been increasing scientific interest in determining how much animals understand, to what degree they are aware of their environment and their own behavior, and to what extent they have humanlike subjective experiences.

The scientists who gathered for the National Zoological Park's second symposium for the public, from which the papers in this volume are derived, presented an exciting overview of the current research and theories relating to animal intelligence, as well as some insight into the historical background of research and speculation on this topic. The twelve chapters of this book explore both learning ability and mental processes in a wide range of animals, from simple worms to rather complex and intellectually sophisticated apes. In each case the reader is confronted, either explicitly or implicitly, with comparisons between the abilities of these animals and those of our own species, *Homo sapiens*. In what ways are we different from each other, and in what ways are we similar?

In the years since Charles Darwin, people have come to accept the fact that the human body is the product of millions of years of evolution, and is quite analogous to the body structure of many other animals. We all see the similarity between, for example, the hand of a human, the wing of a bat, and the flipper of a whale, and we understand that each variation is a solution to a different environmental problem. Brains have also evolved, as has the behavior that they control, but the evolution of mental capabilities as an analogue to the evolution of physical characteristics is not as easily accepted. If animals have subjective experiences that are similar to our own, we would have to seriously reconsider the way we treat them. One of the primary purposes of the symposium was to expose the general public to the latest thinking of scientists who are in the forefront of research on the mental abilities of animals. Armed with such information, people may develop a better understanding of the relationship between humans and nonhuman animals, and thus be better able to make important decisions concerning the other creatures with whom we share our planet.

When we chose the title "Animal Intelligence" for the symposium it was with some misgiving, since psychologists can hardly agree on the meaning of the term "intelligence" when it is applied to humans. It is not unusual to hear a psychologist say, only half joking, that intelligence is what the intelligence tests measure. There is a long-standing debate among psychologists whether intelligence is one unitary and comprehensive mental ability or merely the sum or average of a number of disparate abilities. A recent theory that is gaining popularity claims that there are at least seven distinctly different types of human intelligence (Gardner 1983). It was not unexpected, then, to find that the question of how to define intelligence came up quite often at this symposium and was not always answered in the same way.

In some of the following papers the term "intelligence" is used as more or less equivalent to the ability to learn. This is the sense in which Darwin (1881) used the term when he stated: "If worms have the power of acquiring some notion, however rude, of the shape of an object and of their burrows, as seems to be the case, they deserve to be called intelligent." But the term intelligence can also be used to denote something more, something qualitatively different from such simple learning ability: an internal representation of the outside world, which can then be manipulated symbolically. In other words, a mind. This use of the term intelligence will also be found in several of the following papers. But whatever the specific definition used or implied by a particular author the reader will find a wealth of examples of complex behaviors that have been learned by a wide variety of species. You will meet many fascinating animals: a cat who learned to open a garden gate by grasping the handle and moving the latch, a chimp who was able to tell her trainer that she would like to watch a particular videotape, birds who have learned to help find and rescue people lost and adrift at sea, and even bees who learned to second-guess scientists who kept moving their food supply. But behind the interesting anecdotes there is solid scientific data—and ongoing debate among scientists. Considering the problem of the definition of the term "intelligence," as well as such words as "mind" and "reasoning," perhaps it is ultimately up to the individual reader to decide whether these and related terms can be applied to any particular animals, including humans.

Overview

In the first chapter, David Challinor deals with several different uses of the term intelligence and its applicability to animals. He puts the field of animal behavior in historic perspective and explains the renewed willingness of scientists to use such terms as awareness, thinking, and mind when speaking of animals.

By presenting many fascinating examples of the complex behavior of invertebrates, James and Carol Gould repeatedly deal with the problem of deciding whether any particular behavior is "wired in" by the genes, or involves some degree of intelligence and creativity. They explain different types of learning that various invertebrates and vertebrates show, and analyze the role of innate factors in learning. They present evidence that bees show not only stereotyped forms of "programmed" learning, but also show some rather impressive creative intellectual feats.

Roger Thomas reviews some early laboratory studies that tried to compare directly the learning ability of different animals, and points out some of the difficulties encountered in such attempts. He then presents a basic hierarchy of learning skills and explains how it has been used to evaluate some rather high-level intellectual skills, such as concept formation, in animals, and thus test the relative limits of vertebrate intelligence in properly controlled laboratory experiments.

Duane Rumbaugh and Sue Savage-Rumbaugh take up the question of the evolution of the mental states and cognitive processes that characterize the human mind. They also present strong evidence for the occurrence of very similar states and processes in chimpanzees and other primates. They describe their program to teach chimpanzees a "language" (a form of sign language in this case) and present many fascinating examples of the intelligent use of this communication system by the chimps, as well as many intriguing examples of chimpanzee reasoning, invention, and comprehension of symbols.

William Hodos traces early ideas about the evolution of the brain and the intelligence of animals, and presents recent research findings about the relationship between brain size, brain complexity, and intelligence. He also discusses some of the conceptual problems involved in trying to compare the intelligence of diverse species.

Intelligence, whatever it is, involves responding appropriately to the outside world. Carl Gans emphasizes how animals perceive the world in very different ways, as a result of an amazing variety of different sensory equipment. He explores the variety of mechanisms that different species have evolved to gain information about the world, and focuses on the qualitatively different experiences that different animals can have, even in the same environment. This sensory information, in turn, serves as the basis for all intelligent behavior and mental processes.

Marian Breland Bailey examines the problems involved in assuming that all learning ability is the same and some animals just show more or less of it. She emphasizes the qualitative differences in the types of things that different animals are capable of learning, which usually are related to their natural environments and the specific ecological niche in which the animal evolved. She presents many unusual and often amusing examples of such specialized learning abilities from her long experience in training a remarkable variety of animals.

Colin Beer also takes on the problem of defining intelligence, and presents several different ways that the concept is frequently used by scientists and philosophers—including the idea of artificial intelligence in computers. He discusses intelligence as learning ability and presents some of the costs involved in answering environmental challenges with acquired rather than innate responses. He also emphasizes the important role of complex social structure in stimulating the development of intelligence in primates and other social animals.

Benjamin Beck presents examples of many different types of tool use in animals, some of which reflect a high level of intelligence, some simple trial-and-error learning, and some innate, stereotyped behavior. He compares well-documented examples of tool use by chimpanzees and gulls, and explores the relationship between tool use, cognitive complexity, and internal (mental) representation of environmental features. He also speculates on the early evolution of human intelligence.

Bennett Galef presents examples of apparently complex behaviors that are unique to specific social groups within a species, and traces the history of how the learning of such "traditions" has been interpreted. He discusses the tendency of earlier observers to attribute such behavior to very humanlike patterns of imitation and social learning, and presents recent data that show how complex social patterns in wild rodents can result from much simpler mechanisms.

After presenting a brief historical account of scientific interest in the animal mind, Carolyn Ristau attempts to answer the question which is the title of her paper: Do animals "think"? She presents several lines of reasoning that would support the belief that animals do think, as well as speculation on what they might think about. She then presents a varied set of experiments and field studies, involving many different species, that support the conclusion that many animals act purposefully and experience mental states similar to those that humans experience.

Stephen Vicchio presents an historical account of the interpretation of animal behavior and intelligence in Western culture. He describes how anthropomorphism—the tendency to ascribe human characteristics to animals—has been the most persistent and pervasive method by which people, including scientists and philosophers, have attempted to understand the animal mind. This has sometimes led to such extreme results as putting animals on trial for crimes they have committed. Vicchio's paper closes this volume with both a description of many common misconceptions that people still hold about animal behavior and a plea for viewing all animals objectively as intrinsically valuable beings that, together with our selves and the rest of nature, constitute the earth's grand ecosystem.

Acknowledgments

Funds from the Friends of the National Zoo (FONZ) and an organization that prefers to remain anonymous made possible the symposium and the publication of this volume. We extend our heartfelt thanks to these two organizations. Former National Zoo director, Theodore H. Reed, and current Assistant Director for Research, Devra Kleiman, provided encouragement and support. Michael Morgan and Ilene Ackerman of the Zoo's Office of Public Affairs and intern Sharon Pailen devoted long hours and much energy to supporting the production and coordination of the symposium. And, once again, the FONZ audiovisual operator, Rod Brown, made the symposium a relaxing event for us as he took charge of the technical aspects of each presentation.

Tabetha Carpenter spent many hours on the word processor turning a succession of edited manuscript copies into readable and enjoyable drafts. We are grateful for her devotion to the task. We also thank the symposium moderators Dale Marcellini, National Zoo curator of herpetology, and Devra Kleiman for a job well done in keeping the symposium schedule on time. Finally, public affairs assistant Margie Gibson contributed the drawing of the auroch in chapter 12 while artists Kathleen Spagnolo, Richard Swartz, and Vichai Malikul provided remarkable illustrations for all of the chapters.

Select Bibliography

Darwin, C. R. 1881. *The formation of vegetable mould through the action of worms.* London: J. Murray, p. 97.

Gardner, H. 1983. *Frames of mind: The theory of multiple intelligences.* New York: Basic Books.

Introduction:
The Issue of Animal Intelligence

David Challinor

The papers in this volume all relate in some way to the concept of intelligence and how animals exhibit intelligence. But, as many of the authors will point out, the term intelligence is imprecise, especially when applied to animals. It is a quality difficult to measure directly, an abstraction that can be grasped only by its effects.

The dictionary defines intelligence as "the capacity to acquire and apply knowledge," and knowledge is defined as "familiarity or understanding gained through experience." I perceive the word intelligence in a slightly different way. For me, it implies, first of all, self-awareness as a physical and social being. An intelligent being possesses thoughts and actions with moral and social values, a complex of thoughts and images from the past, present, and future, and perhaps the ability to communicate complex thoughts by some verbal or nonverbal linguistic ability. When I think of these attributes, I realize that I have given a rough definition of human intelligence—and thereby have shown my own bias (as well as, I would venture, the biases of many people).

The word intelligence is freighted with human feelings and emotions, and thus, is not easily applied to nonhumans. That may be why, during the first decades of this century, scientists working in the then new field of animal behavior endeavored to legitimize their discipline by avoiding the use of terms such as intelligence, thought, awareness, mind, or consciousness when they were referring to animals. They concentrated instead on what could be directly observed, scientifically tested, and statistically validated. Many of us remember the experiments emphasizing simple behaviors: pigeons pecking keys, rats pressing levers, cats opening doors, etc. Human-biased words and concepts became taboo among many of these professionals because they were too "unscientific" to warrant study. But today one hears such words as "intelligence," "mind," "thinking," and "awareness" applied to many animals. What has changed to remove the taboo on the use of these words in studying animal behavior?

The field of animal behavior has moved out of its infancy. The research of the past twenty or thirty years has added new insights into animal behavior, so that its study is now a secure and respected discipline. Techniques have become increasingly sophisticated, and the field of animal behavior is now a legitimate science. Although it is unlikely we will ever get inside an animal's mind, any more than we can ever penetrate the mind of a fellow human, we are beginning to acquire a clearer picture of how animals "tick," for lack of a better word.

A basic issue of the whole question is whether the term intelligence as applied to animals is useful. This is obviously a semantic problem, perhaps almost a personal one. I propose the term is worth using as an umbrella term—a category or a chapter heading that could be subdivided into more meaningful subcategories. Many of the following chapters will do just that. Investigators working within the boundaries of animal intelligence may focus on such specific topics as learning, instinct, intention, choice, self-awareness, and deception—topics that may lend themselves, in varying degrees, to scientific testing. But more general interpretations, perhaps even cautious speculations, are often derived from such research. This is the second issue with which we are confronted. Can we accept interpretations and cautious speculations about animal intelligence as valid? The answer depends, of course, on how solid the facts are on which such speculation is based. I think you will find that the research reported in this volume, to the extent possible, reflects this essential kind of factual underpinning.

A landmark in legitimizing expansion beyond the early conservative position of animal behaviorists was the 1981 Dalhem Conference, "Animal Mind/Human Mind," which was held in West Berlin. This conference was attended by well-known psychologists, zoologists, physiologists, and other experts with an interest in exploring new frontiers in animal behavior, and the papers given there have spurred further professional interest into the nature of the animal mind. Just as the Dalhem Conference made acceptable the scientific investigation of such topics as mind, awareness, consciousness, thinking, and intentions in animals, it is my hope that these symposium proceedings will convince the interested layman that the study of animal intelligence is now a legitimate scientific pursuit, and no longer the province of philosophy, fantasy, or unfounded speculation.

Invertebrate Intelligence

James L. Gould and Carol Grant Gould

The females of one species of digger wasp maintain simultaneously several burrows containing developing offspring. As the day begins, the female visits each of the concealed tunnels and checks on her young. On the basis of this round of inspection, she knows which burrows still contain eggs that require no food, which are occupied by young larvae that will need two or three caterpillars to eat, which have older larvae that will need many more caterpillars, and which contain newly pupated offspring that need to be sealed off to complete their development. On the basis of her visit to these five or ten different subterranean nurseries, the wasp knows how much prey to capture and where to take the spoils of her hunting trips. On the surface, this behavior looks both thoughtful and intellectually demanding. But how are we to know what is going on in the mind of the wasp? Are invertebrates something more than mere mindless circuitry, as we presume ourselves to be, or are they merely nervous, six-legged computers? Since the mind is, by nature, a private organ, how are we to judge from an animal's overt behavior whether we are observing a well-oiled machine or a creature with some degree of intelligence and creativity? Particularly with insects whose chitinous exoskeletons make it difficult to consider them in anthropomorphic terms, how are we to discover the extent to which they might be acting intelligently?

Criteria for Attributing Intelligence to Animals

Several lines of evidence for animal intelligence have come to the fore, but a little careful thinking, observation, and experimentation indicate that most of these criteria are untrustworthy. One intuitively powerful argument, for instance, is that since animals regularly face problems and solve them in sensible ways, they must have some intellectual grasp of the problem. When a honey bee, for instance, encounters a dead bee in the hive, it very properly tosses it out of the colony. But experience tells us that adaptive behavior most often reflects the intelligence of evolution rather than of the animals it has so carefully programmed. Bees recognize their dead colleagues by means of a "sign stimulus"—a single key feature of a stimulus object that is taken to represent the entire object. In this case a special "death odor," possibly oleic acid, triggers the removal behavior. So mindless is the "wiring" of this proper exercise in colony hygiene that a drop of oleic acid on an otherwise innocuous piece of wood or even on a live bee results in the removal of the offending object. The sight of one bee carrying out a struggling sister or even the queen should convince us that behavior can seem intelligent in its normal context without any need for the intellectual participation of the actors.

A second criterion frequently suggested is that the very regularity and invariability of such robotlike behavior may be a guide to what behavior is performed automatically and without the need for thinking. As we all know from personal experience, intellect will often come up with two very different solutions to the same problem in two different individuals, or even in the same individual on two different occasions. An automatic computer would come up with one "best" answer. When the nineteenth-century French naturalist Jean Henri Fabre interfered with the prey-capture ritual of a cricket-hunting wasp by moving its paralyzed victim, he stumbled upon some of the wiring that runs the wasp's routine. The wasp, whose behavior appears eccentric but intelligent, invariably leaves the cricket she has caught lying on its back, its antennae just touching the tunnel entrance, while she inspects her burrow. Each time Fabre moved it even slightly away from the entrance, the reemerging wasp insisted on precisely repositioning the cricket and inspecting the tunnel again. Fabre continued this trivial alteration 40 times and the wasp, locked in a behavioral "do-loop," never thought to skip an obviously pointless step in her program. Clearly the wasp is a machine in this context, entirely inflexible in her behavior.

The remarkable persistence of the wasp's performance serves also to remind us that most other animals have contingency plans to extricate them from such behavioral culs de sac. By far the most common escape mechanism for organisms ranging from bacteria to human beings is "habituation," a kind of behavioral boredom by which an animal becomes less responsive as it repeatedly encounters the same stimulus. But habituation and other such escape strategies are not the result of active intellect. They are merely sophisticated programming ploys, and in the sea slug *Aplysia*, the neural and biochemical bases of the machinery are pretty well understood. In this fascinating marine mollusc the sensory receptors around its delicate siphon and gill are wired to a set of motor neurons which, when they are stimulated, cause a group of muscles to pull this underwater breathing apparatus back inside the body for safety. But what constitutes a potentially dangerous stimulus in calm water is quite different from what is appropriate in a rough current. The connections (called synapses) between the sensory cells and the motor neurons adjust themselves to the background level of stimulation so that only something really out of the ordinary causes them to fire. Habituation, then, though highly adaptive, is an automatic, intellectually undemanding process that depends on nothing more mysterious and metaphysical than the adjustment of particular molecular gates in the synaptic membrane. The mindlessness of this acquired behavioral numbness is illustrated by the contrary phenomenon of sensitization: almost any irrelevant but novel stimulus can instantly destroy habituation's insensitivity by activating a compensatory set of chemical channels.

Other sorts of seemingly intelligent behavioral variability though, cannot be accounted for either by "noise" in the computer or by habituation. Honey bees, for example, show spontaneous preferences for certain colors and shapes of artificial flowers, with many-petalled violet flowers being the most attractive to the apian mind. This display of aesthetic preference is not absolute, but instead probabilistic: given a choice between two colors that we know from learning experiments they can distinguish reliably—violet and blue, for example—the bees will choose their favorite, violet, only 70 percent of the time rather than 100 percent. Similarly, in a conflict situation an animal will sometimes fight and sometimes flee. Even in experiments in which care has been taken to factor out the role of immediate past experience, this sort of predictable variability persists.

Can the perplexing unreliability of animal behavior be taken as evidence for something more (or perhaps less) than machinery making decisions? Probably not. Game theory demonstrates that it is usually most adaptive to be variable or unpredictable, so long as evolution or personal experience takes care to set the odds appropriately. Though flowers may more often be violet than blue, it makes sense to try blue-colored objects from time to time rather than to concentrate exclusively on violet. Even this sort of quasi-aesthetic "decision" makes enough evolutionary sense that there is a good chance that it results from programming rather than intelligence; and, indeed, in most carefully studied cases it is clear that variability *is* innate.

But though a great deal may be programmed into animals, there must surely be a limit to the complexity possible. There must be a point beyond which no set of onboard computer-like elements can suffice to account for an animal's apparent grasp of its situation, particularly in the face of variable or unpredictable environmental contingencies. The difficulty in drawing this intellectual line, however, is daunting. Some of the most impressively complex examples of behavior we see are known to be wholly innate. In total darkness, without prior experience, and with the location of potential anchor points for the support structure unpredictable, a mere spider sets about constructing a precise orb web, a complex network of several different kinds of threads held together with hundreds of precisely placed "welds." Even damage during construction is automatically repaired. All this is accomplished through one master program and several subroutines and requires no conscious grasp of the problem.

So too it is with the famous dance "language" of bees. When a forager returns from a good source of food, she performs a dance that indicates to other bees the distance, direction, and quality of the food she has discovered. The direction is encoded by transferring the horizontal coordinates outside the hive to a vertical sheet of comb in the dark hive, defining the sun's azimuth as "up" on the comb, and orienting the dance the appropriate number of degrees to the right or left of vertical, thereby matching the angle outdoors between the food source and the sun (figure 1). The distance is encoded by the number of body "waggles" in the dance so that each wag corresponds to about 50 meters for the German race, 25 meters in the Italian dialect, but only 15 meters in the Egyptian honey bee's version of the dance (figure 2). But though this abstract, symbolic communication system is second only in complexity and information capacity to hu-

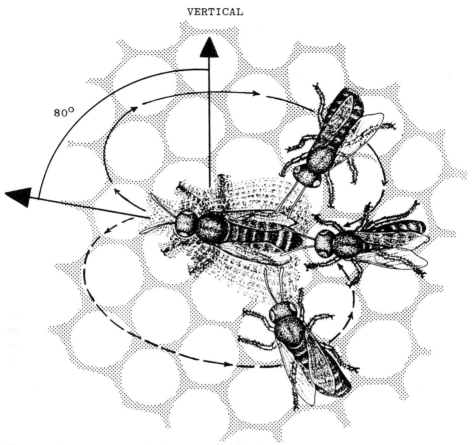

VERTICAL

80°

Figure 1 Foragers report the location of a food source by means of a waggle dance performed in the shape of a figure-8 on the vertical sheets of comb in the hive. During the straight-run portion of the dance (where the two lobes of the figure-8 overlap), the dancer vibrates her body from side to side (shown here by shaded silhouettes). The angle of the straight run relative to vertical corresponds to the angle of the food source relative to the sun. In the case illustrated here, the food is 80° to the left of the sun. The number of waggles in the straight run indicates the distance to the food (see figure 2).

man speech, it is wholly innate: bees can both perform and understand dances with no previous experience.

The use of a "subroutine" strategy for dealing with more difficult and less predictable behavioral challenges is especially obvious in navigation. Honey bees, for example, regularly use the sun as their compass, compensating for its changing azimuth as it moves

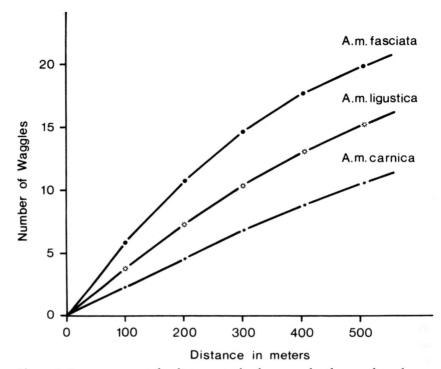

Figure 2 Foragers report the distance to food sources by the number of waggles in the straight run. Different races of bees, however, have different distance dialects. The German race, *carnica*, for instance, values each waggle at about 50 meters, whereas the Italian race, *ligustica*, converts distance into waggles at approximately 25 meters per waggle. The Egyptian race, *fasciata*, has the shortest dialect at 15 meters per waggle.

from east to west. This is a formidable task even for a human navigator, but as we have found out in the past few years, the bees' trigonometric adjustments are perfectly mindless, depending only on a memory of the sun's azimuth relative to the bee's goal on the previous trip (or day) and an extrapolation of the sun's current rate of azimuth movement. The strategy is innate, though the program must include steps for measuring the relevant variables when necessary. Bees recognize sun by an equally innate criterion—its low ratio of ultraviolet to visible light—so that a dim 10°, triangular, highly polarized green object against a dark background is just as acceptable as the actual, intensely bright ½° circular, unpolarized white sun that the bees see almost every day in the normal sky.

When the sun is removed (obscured perhaps by a cloud, a land-mark, or the horizon) the bees' whole sun-centered system is discarded in favor of a backup system—a separate navigational subroutine—based on the patterns of polarized light generated in the sky by the scattering of sunlight. This analysis itself is composed of a primary and backup system and uses sign stimuli and very simple processing. When the polarization is removed as well (as on overcast days, for example), bees fall back on yet a third system, based on landmarks, and there is no reason to suppose we have exhausted the set of failsafe plans built into bees. The apparent complexity of the formidable navigational behavior that many insects display is, in fact, based on the interplay of groups of subroutines that are themselves quite simple. They depend as a rule on the same sorts of schematic stimulus-recognition systems and simple processing seen in less elaborate behavior. Since a staggering degree of behavioral complexity can be generated by a set of individually simple subroutines, mere complexity of behavior cannot be in itself a trustworthy guide to intelligence.

Another commonly accepted indication of intelligence is the way animals deal with the unpredictable contingencies of their world through learning; and it is here that our intuition tells us that we must be dealing with something very like intellect. After all, learning suggests to most of us some degree of understanding, some conscious comprehension of the problem to be solved. Alas, headless flies can learn to hold their legs in a particular position to avoid a shock and even solve the problem faster than those still encumbered with brains.

Learning theory has traditionally recognized two general sorts of learning: associative learning (also known as classical or Pavlovian conditioning) and trial-and-error learning (also known as operant or Skinnerian conditioning). Associative learning is the process by which an animal comes to replace an innately recognized cue—the sign stimulus—with another cue or set of cues. It is nature's version of inductive reasoning: animals learn only cues that tend to *predict* the imminent arrival of something desirable, like food, and the reliability of the new cue does not need to be by any means perfect. But we need not imagine that the animal being educated actually comprehends the problem. The learning appears to be perfectly mechanical, as indicated by clear, species-specific rules. For instance, if two novel stimuli—a tone and a light, say—precede the sign stimulus and occur simultaneously, the animal, whether a garden slug or a rat, will learn both and be able in the future to use either alone. But if, on the other hand, the tone precedes the light, the animal does not now have the wit to see that the tone is as good a predictor as the light. Instead, it learns only that the light is associated with the sign stimulus under these circumstances. So too it is that many animals have strong, context-specific cue biases: a chicken will quickly learn to associate a light but not a tone with food, but can associate the same tone (though not the light) with impending danger. Such examples are almost endless.

Trial-and-error learning, on the other hand, involves learning to perform a novel motor behavior that is used to solve a problem posed by nature. Animals discover by experimentation what works and what does not, and so experience shapes the behavior. This process is in many ways analogous to deductive reasoning, but again there is no reason to suppose that the learning requires any active intelligence on the part of the "student." Relatively rigid mental rules underlie trial-and-error learning, which animals appear unable to overcome even when it seems obvious that they are poorly matched to the situation. For instance, a rat being trained through trial-and-error to avoid a shock will take a long time to learn that pressing a bar will prevent this aversive stimulus, if it ever learns at all. This is not a consequence of any generalized reluctance to press bars—indeed, bar-pressing is the most common behavior to be conditioned when a food reward is involved. Similarly, a task in which the animal must run to the other end of a box to avoid a shock and then back again to avoid the next (so that no particular end is ever consistently safe) is very difficult for a rat to comprehend, though it is easy to master when food is involved. Wheel-running is easier to condition as a shock-escape behavior, while leaping out of the box is quickest of all. Animals exhibit many such "biases" or "predispositions" that illustrate that the ease or difficulty of a learning task depends on the context and the innately specified rules for dealing with it.

The Role of Learning in the Life of Honey Bees

These problems and anomalies, however, only seem to turn up in the laboratory. In the real world, learning, however rote, works smoothly and efficiently as an animal deals with the challenges presented by its environment. The roles of inductive and deductive learning in the lives of animals and how these processes relate to the issue of intelligence may be illustrated by how they work in honey bees. As we shall see, there is as much in the organization of bee learning as in that of rats to suggest the gears and wheels of an automatic pilot rather than any aware intelligence. When a honey bee discovers a flower, for example, she sets in motion a learning sequence that seems utterly mechanical. A forager learns many things about a food source that will aid her in the future, including its color, shape, odor, location, nearby landmarks, time of nectar production, how to approach, land, enter and reach the nectar, and so on. Color learning, for instance, has all the marks of associative learning: bees have an innate program that recognizes flowers by their dark centers and light petals (as seen in ultraviolet light—these markings are usually invisible to our eyes). After it has served its purpose, though, this sign stimulus is replaced through associative learning with a far more detailed picture of the flower. Color is learned only in the final three seconds as the bee lands: the color visible to the bee before the landing sequence and the colors it sees while standing on the flower to feed and while circling the blossom before flying off simply never register. Experimenters can change them at will and the bee will never be fooled. A naive bee *carried* to the feeder from the hive and placed on the food source will circle repeatedly after taking on a load of sugar water as if "studying" the source, but when she returns a few minutes later she will be unable to choose the correct feeder color. And yet, so mechanical is this learning routine that if we interrupt such a bee while she is feeding so that she must take off and land again on her own accord, that second landing permits her to choose the correct feeder color on her next visit. Similarly, bees learn landmarks after taking off: a recruit who arrived and fed at the feeder, but was transported back to the hive while feeding, returns to the hive without the slightest memory of the landmarks she must certainly have seen on her arrival.

Other aspects of the associative component of flower learning seem equally curious. Although a bee learns a flower's odor almost perfectly on one visit, she must make several trips to learn its color with precision; and even then a bee never chooses the correct color 100 percent of the time. Shape is learned less quickly, and time of day more slowly still. It is as though perfection, clearly possible in other contexts, is not in this case desirable. It appears, in fact, that the speed and reliability of a bee's flower memory at least roughly corresponds to the degree of variability it is likely to encounter among flowers of the same species in nature (including variation from day to day of an individual blossom). In fact, the rate at which a bee learns each component differs dramatically between various geographic races of honey bees, strongly implicating a genetic basis for the different learning curves.

Once a bee has learned how to recognize a particular kind of flower and when and where to find it, it is as though the information is stored in an appointment book. As a result, a bee cannot learn that either of two flowers provides food at one time of day. Moreover, changing any component of the set—the odor, say, which is learned to virtual perfection after one visit—forces the bees to relearn painstakingly all the other pieces of information at their characteristic (slower) rates even though they have not changed. So, logical and impressive as the associative flower learning of honey bees appears, it seems clear that these hard-working insects are simply well-programmed learning machines, attending only to the cues deemed salient by evolution (and then only in well-defined contexts and often during precise critical periods) and filing the information thus obtained in preexisting arrays. Nothing in this behavior, wonderful as it is, suggests true flexibility or awareness. Nor is the situation any different when we look at the trial-and-error component of the behavior by which bees learn how to harvest a flower species efficiently.

Cognitive Processes in Insects

We can see that the widespread strategy of programmed learning is the means by which the genes tell their dimwitted couriers when and what to learn (how else could an insect reason it out?) and then what to do with the knowledge thus obtained. Even the impressive case of the hunting wasp described earlier can be revealed, with a little delicate manipulation, to be devoid of real intelligence. If the occupants of the various burrows are switched during the night, their mother has no difficulty in adjusting to the changes and stocking each nursery appropriately the next morning. If, however, the wasp's offspring are exchanged *after* her morning inspection round, the wasp, oblivious to the fact that the burrow now has an egg instead, will spend the day stocking with caterpillars a cell that earlier had an older larva. Indeed, the wasp will probably touch and even appear to examine the egg many times over the course of the day without the inappropriateness of her behavior sinking in. Similarly, a young larva will be sealed off to starve because earlier the tunnel had contained a pupa.

Though the wasp shows no glimmer of insight, there *are* cases of apparently self-directed learning that may admit to another explanation. Indeed, one of the major factors that led to the demise of classical behaviorism was the discovery that animals can learn a motor behavior—which way to run in a maze to get to some food— without the need for either associative learning or overt trial-and-error experimentation. A rat, for instance, carried passively to each of two "goal boxes" at opposite ends of a runway and shown that one contains food and the other does not will, when released, run unerringly to the box with food. This phenomenon, which we call "cognitive trial and error," requires a deductive process to go on inside the mind of the animal without its actually *trying* different behaviors. The animal, be it the rat in its maze or a chimpanzee stacking boxes to reach a bunch of bananas overhead, must reason out a course of action in its mind. Here is something that seems very like intelligence, and we must ask whether it is really that or merely another clever but mechanical programming finesse that we do not yet see.

There are among honey bees three reported examples that appear at first glance to qualify as cognitive trial and error. The first concerns their avoidance of alfalfa (lucerne). These flowers possess spring-loaded anthers that give honey bees a rough blow when they enter. Although bumble bees have evolved as pollinators of alfalfa and do not seem to mind, honey bees, once so treated, avoid alfalfa religiously. Placed in the middle of a field of alfalfa, foraging bees will fly tremendous distances to find alternative sources of food. Modern agricultural practices and the finite flight range of honey bees, however, often bring bees to a grim choice between foraging alfalfa or starving.

In the face of potential starvation, honey bees finally begin foraging on alfalfa, but they learn to avoid being clubbed. Some bees come to recognize tripped from untripped flowers and frequent only the former, while others learn to chew a hole in the side of the flower so as to rob untripped blossoms without ever venturing inside. Who has analyzed and solved this problem—evolution, or the bees themselves? It may be that both cases are standard, prewired back-up ploys: differentiating tripped from untripped flowers could be simply a far more precise use of the associative learning program, while chewing through may be a strategy normally held in reserve for robbing flowers too small to enter.

Another slightly eerie case is not so easy to dismiss. While training bees to fly to an artificial food source, we systematically move the food farther and farther away on successive trials. There comes a point at which some of the trained foragers begin to "catch on" and *anticipate* subsequent moves and to wait for the feeder at the presumptive new location. This seems an impressive intellectual feat. It is not easy to imagine anything in the behavior of natural flowers that might have caused evolution to program bees to anticipate regular changes in distance.

Along the same lines, we have seen behavior on several occasions during experiments on bee navigation that appears to reflect an ability to form what experimental psychologists refer to as a "cognitive map." The classic example of this phenomenon is the ability of a rat to explore each arm of an eight-armed maze in a random order without inspecting any arm twice (Olton 1978). Although the rat can be fooled if the maze is rotated, it does have a long-term spatial memory. When the rat is removed from the maze after having explored only five arms and returned to the apparatus several hours later, it will first examine the three arms it did not have the time to check earlier. This ability to form a mental map and then formulate appropriate behavior (perhaps by imagining various alternative scenarios) seems very similar to the ability of chimpanzees to imagine the solution to the hanging banana problem by stacking the boxes in their minds before actually performing the behavior; and, of course, the same process goes on in our own minds all the time.

The first hint of such an ability in bees came years ago when von Frisch discovered that bees that had flown an indirect route to a food source were nevertheless able to indicate by their famous communication dances the straightline direction to the food. By itself, it is easy to interpret this ability as some sort of mindless, automatic exercise in trigonometry. Three years ago we trained foragers along a lake and tricked them into dancing to indicate to potential recruit bees in the hive that the food was in the middle of the lake. Recruits refused to search for these food sources, even when we put a food source in a boat in the lake at the indicated spot. At first we thought that the recruits might simply be suffering from some sort of apian hydrophobia, but when we increased the distance of the station so that the dances indicated the far side of the lake, recruits turned up on the opposite shore in great numbers (figure 3). Apparently they "knew" how wide the lake was and so were able to distinguish between sources allegedly in the lake and sources on the shore. We see no way to account for this behavior on the basis of either associative

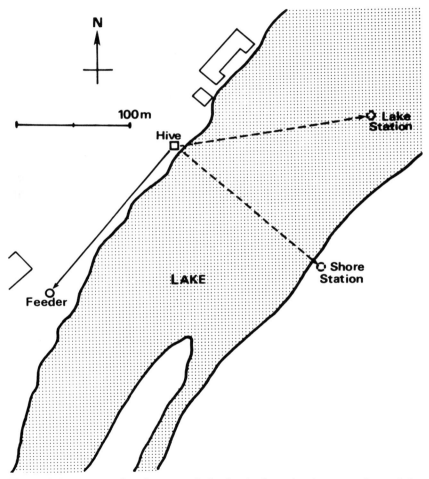

Figure 3 Recruits refused to search for food when the dances indicated the lake station, but responded in large numbers when the dances directed them to a spot at the same distance from the hive but on the far side of the lake.

or trial-and-error learning. This ability is accounted for most simply if we assume that the recruits have mental maps of the surroundings on which they somehow place the spots indicated by the dances.

This interpretation is further reinforced by another observation. While exploring the question of whether bees can use direction information gathered on the flight back from the food, we transported foragers (caught as they were leaving the hive for natural sources) to an artificial feeder in the middle of a large parking lot hundreds of meters from the hive. After being allowed to feed, the majority of foragers circled the feeder and, in many cases, departed directly for the hive (which was out of sight). Most of the young bees, however, circled helplessly and never got home. When the successful foragers arrived at the hive, many danced to indicate the parking lot. Now for a bee to know the location of a barren parking lot that had certainly not been on their list of flower sites, it seems most reasonable to suppose that they were able to place it on some sort of internal map and then work out the direction home. That only older (and presumably more experienced) bees were successful at this task is consistent with this interpretation.

Taking these cases at face value, does the apparent ability to make and use maps provide convincing evidence of active intelligence? And if so, why are bees so thoroughly mindless in other contexts? Speculation on the second question is easier than on the first. Intuitively, it seems reasonable to suppose that if we were designing an animal we would "hard-wire" as much of its behavior as possible. Where there is a best way of doing something, or finding out how to do something, there seems little point in forcing an animal into the time-consuming, mistake-prone, and potentially fatal route of trial-and-error learning. But where explicit programming will not serve, it seems equally reasonable to direct an organism to fall back on "thinking," particularly when the solution to a problem can then be wired into the system for later service. Much of human behavior seems to fall into this neurologically economical pattern: we work hard to master a problem, then turn the solution into a mindless, rote unit of behavior. Difficult problems like learning to type, ride a bicycle, tie shoes, or knit seem almost impossible at first, but once learned become as matter-of-fact as breathing or walking.

But whether cognitive trial and error qualifies as intelligence is more difficult. On the one hand, we can imagine how we might go about prewiring a Cartesian map and how we could then encode the instructions by which the information to fill the map should be gathered, stored, and used. On the other hand, there is increasing evidence that many of the intellectual feats of our own species—language acquisition, Aristotelian logic, categorization, pattern recognition, and the like—are themselves based on preexisting wiring and storage. The more we learn about the brain, the more clearly we see how its specialized wiring affects what we are and how we think. It may be that the question is one of degree: to what extent is a pocket calculator "intelligent"? Does a TI-59 calculator with its hardwired navigation module installed—a good approximation of a small part of a honey bee brain—qualify? What about a chess-playing machine, programmed to examine the board and then "imagine" thousands of possible moves and evaluate them in relation to each other? Or is it the provision for automatic self-programming such as we see when a flower trains a bee to exploit it that is intelligence? Does the apparent existence of preordained storage and processing arrays, as is strongly suspected for honey bee flower memory and human language, by itself disqualify a behavior using such innate neural specialization from the realm of intelligence?

Conclusions

The more we look at the behavior of insects, birds and mammals, including man, the more we see a continuum of complexity rather than any dramatic difference in kind that might separate the intellectual Valhalla of our species from the apparently mindless computations of insects. We see the same biochemical processes, the same use of sign stimuli and programmed learning (even in language acquisition), identical strategies of information processing and storage, the same potential for well-defined cognitive thinking, but very different storage and sorting capacities and, most of all, very different intellectual needs imposed by each species' niche. We are inclined to the paradoxical view that, on the one hand, cognitive trial and error is the *sine qua non* of animal intelligence—and, in the case of our species, its synergistic combination with symbolic language (which permits us to use abstract concepts in these mental experiments) has, for better or worse, vastly increased our intellectual potential— while, on the other hand, this very strategy, which carries animals at least temporarily out of many of their genetic constraints, is itself carefully prewired into the minds of the species for whom it is adaptive. We suspect that the mental game board, the cognitive players, and the rules for moving the pieces are as carefully preordained and tailored to the needs of an animal as are its learning programs, its innate motor behavior, and its specialized physiology. If this view is correct (and perhaps even if it is not), the very limited but easily studied thinking of insects may offer a marvelous opportunity to study and understand the essence of intelligence and suggest how our own awesome mental powers are organized and have evolved.

Select Bibliography

Gould, J. L. 1982. *Ethology: The mechanisms and evolution of behavior.* New York: W. W. Norton.

Gould, J. L., and C. G. Gould. 1981. The instinct to learn. *Science 81* 2(4): 44–50.

———. 1982. The insect mind: Physics or metaphysics? In *Animal mind— Human mind*, ed. D. R. Griffin, 269–98. Berlin: Springer-Verlag.

Olton, D. S. 1978. Characteristics of spatial memory. In *Cognitive processes in animal behavior*, ed. S. H. Hulse, H. Fowler, and W. K. Honig. Hillsdale, N.J.: Lawrence, Erlbaum Assoc., Inc.

Vertebrate Intelligence:
A Review of the Laboratory Research

Roger K. Thomas

Are dogs smarter than cats? Are pigs smarter than horses? Are dolphins smarter than apes? Questions like these have been asked for centuries, but regrettably, I will not be able to answer them. What I will try to do instead is explain why such questions have not been answered despite many attempts to do so and suggest some ways in which such questions might be answered. In the course of doing this, however, I will be describing some interesting intellectual achievements of a variety of vertebrate species.

The Concept of Intelligence

Intelligence is a fictitious entity. It has no physical existence. No structure in the brain or elsewhere corresponds to it. No standard definition of intelligence exists, and the concept means different things to different people. As a subject for scientific study, therefore, there are certainly easier topics to tackle. But, if we want to know about the intelligence of nonhuman animals, we must first try to agree on a definition of intelligence, or at least on a framework within which to consider it. And if we wish to compare the intelligence of different animals, we must find a scale of measurement or "common denominator" that is suitable for comparing species as diverse as those that inhabit the Earth.

Some have argued that this cannot be done, that there is no common denominator because species vary too widely. Such people are likely to view intelligence as an abstraction that represents how well each species adapts and survives in its environment. According to this view, there are many kinds of intelligence, and it is not meaningful to compare them. One cannot refute this viewpoint; one can only decide whether to accept, reject, or be willing to compromise its premises. If one prefers to accept its premises without compromising, then we may as well admit that the cockroach is as intelligent as the human and let it go at that. However, if we are willing to compromise, then we may persist with the notion that a common denominator can be found with which to compare animal intelligence.

Intelligence, as I see it, is closely related to adaptability and survival. Some aspects of intelligence are genetically determined, while other aspects involve learning. Some species' survival depends primarily, if not exclusively, on genetically determined behavior, but other species depend to varying degrees on what and how well they learn. Comparing the inherited components of intelligence does not seem to be appropriate, or, at least at this time, there does not appear to be a meaningful basis for comparison. Comparing learning abilities, however, is reasonable and feasible, provided we do not mismeasure them—and the possibility for mismeasurement is considerable. For the remainder of this paper, intelligence will be treated as being equivalent to *learning ability*, and differences in intelligence will be regarded as differences in learning ability. This is not to deny that there are components of intelligence other than learning ability, but rather to simply acknowledge that, at present, there is no way to measure and compare these other components.

The Mismeasure of Learning Ability

Learning ability is usually determined in the laboratory by training an animal to perform some task and keeping a record of the number of trials required to master the task (usually referred to as the number of "trials to criterion") as well as the number of errors made during the process of learning the task. It is important to emphasize that such scores are measures of performance, and that a distinction must be made between learning and performance.

While learning ability certainly affects performance, other factors also influence performance but bear little relation to learning ability or intelligence. This point is especially relevant in a comparison of species. For example, as Carl Gans (this volume) explains, sensory abilities of species are rarely equivalent: some animals see color while others do not, some animals have senses of smell or hearing superior to others, and so on. Such sensory differences might give one animal an advantage over another in the performance of a task without reflecting a difference in learning ability. Animals also differ in their motor abilities, that is, their abilities to make the response that the learning task requires. Again, such motor differences might affect performance without reflecting a difference in intelligence. A third factor is motivational differences: one animal may be motivated to perform well while another animal may be less motivated. And these do not exhaust the variables that, alone or in combination, might affect performance and therefore *contaminate* or *confound* the assessment of learning ability or intelligence.

Because performance differences may be influenced by such variables, the general view in recent years has been that quantitative differences such as the number of trials to criterion, or the number of errors made, should not be used to compare intelligence between species. A better approach is to ask whether an animal can learn a task when conditions are suitable—for example, when the task is appropriate in terms of the animal's sensory and motor abilities, and when the animal is properly motivated. If the animal *can* master such a task, then the ability to learn that task is within the animal's intellectual capacity. To compare species, one should use a series of tasks that examine different capacities. If the series of tasks represents a meaningful hierarchy of abilities, then it might be possible that the subject that succeeds further up the hierarchy is more intelligent.

The points presented so far were usually overlooked in early laboratory studies that attempted to compare the intelligence of different species of vertebrates. But even when they were not overlooked, and investigators tried to equate testing conditions, they could never be sure that the testing conditions were, in fact, equal. Among the most interesting of these early studies is one by Harold Fink, which was published as a monograph in 1954 as *Mind and Performance.* I feel apologetic for choosing this study to illustrate the mismeasure of learning abilities because in many ways it was a monumental and heroic effort, and I respect and admire what Dr. Fink attempted to do. Despite the flaws that I will point out in the study, many interesting and useful data are to be derived from it.

Fink constructed what he called the Arrow Maze (figure 1) for the purpose of comparing the learning abilities of reptiles, birds, and mammals. The animals were trained initially to find a food reward at the end of alley 1. When the animal had done 10 successive trials without error, it was then required to learn to find the reward at the end of alley 2. When the criterion of 10 successive errorless trials was reached in alley 2, it was required to learn to go down alley 3. Finally, it was required to learn to go down alley 4. Each animal learned to go to each goal in this same sequence, and a record was kept of the number of trials to criterion and errors on each goal.

Two factors made Fink's study especially interesting. First, he used 50 humans (college students), 9 pigs, 10 dogs, 10 cats, 10 chickens, 20 rats, 1 goat, 1 rabbit, and 45 turtles and tortoises. No other study to my knowledge has compared such a variety of subjects on the same task. Humans were blindfolded and tested on a version of the Arrow Maze that could be traced with a finger, and unlike the animals, they were not given a food reward but were merely told when the made an error or a correct response. The second interesting feature is that Fink acknowledged the need to equate testing conditions for each species and made a considerable effort to do so.

To equate for running speed, Fink measured the running speed of animals from each group by having them run down a straight alley for a food reward. As might be expected, the turtles ran the slowest (5 yards/minute on the average). Using the turtle's running speed as a reference point, Fink determined how much time each species should be allowed to perform each trial. The turtles were allowed 30

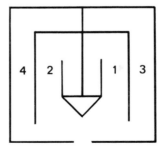

Figure 1 The Arrow Maze was constructed by Fink to compare the learning abilities of reptiles, birds, and mammals. Subjects used in the study included turtles, chickens, rats, dogs, cats, pigs, and college students. (Illustration redrawn by Kathleen Spagnolo from Fink 1954.)

minutes per trial. Most of the animals were allowed from 2 to 5 minutes, with dogs being given the least time at 1.3 minutes. However, while this was an admirable attempt to compensate for different running speeds in evaluating the success of each trial, it created another problem. One of the oldest known principles of learning is that of contiguity, that is, the relationship between two events is more likely to be learned if they occur closely in time or space. The animal that gets to the goal fastest presumably has a time advantage in associating its path to the goal with its discovery of the reward in the goal.

Fink tried to equate motivation by feeding the animal its total daily allotment of food as rewards in the training trials. He adjusted the number of trials per day and the amount of food on each trial based on his impression of how hungry or satiated the animal appeared to be. While this procedure does correct somewhat for differential food requirements, it does not insure equality of motivation. If anything, it may have been counterproductive, because the normal meal patterns of animals vary. Some animals, such as rats, eat small amounts continuously throughout the day, while others eat most of their food at one time, as dogs do, or consume several distinct meals each day, much as humans do. Thus Fink's approach may have approximated the normal feeding pattern of some of his subjects, but it would have been disruptive of others'.

There is also the question of the particular foods given to each species. Naturally, it was necessary to choose appropriate foods for each species, but, as Fink acknowledged, one could not assume that the foods were equally preferred. He also acknowledged the problem of differences in sensory abilities, but he did not even attempt to control for such differences. Thus, on the one hand, Fink viewed the animal's performance, without regard to its physical advantages or disadvantages, as the appropriate test of ability, and on the other hand, he tried to equalize certain factors such as running speed and motivation. In any event, let us see what he found.

Fink used three measures of learning: the number of trials it took to learn the sequence of goals, the number of trials on which an error was made, and the number of alley-entrance errors. Table 1 shows the species in ranked order on the three measures, as well as the overall score based on the average of the other three. Despite the slight variations in rank order, there is about 90 percent agreement among the measures.

Table 1 Rank Order of Performance in Fink's Arrow Maze

Total Trials	Error Trials	Errors	Overall
Human	Human	Human	Human
Opossum	Opossum	Opossum	Opossum
Pig	Pig	Dog	Pig
Dog	Dog	Pig	Dog
Chick	Eastern Painted Turtle	Goat	Eastern Painted Turtle
Eastern Painted Turtle	Rat	Eastern Painted Turtle	Goat
Rat	Chick	Rat	Chick
Goat	Goat	Rabbit	Rat
Cat	Rabbit	Chick	Rabbit
Rabbit	Cat	Cat	Cat
Other Turtles and Tortoises	Other Turtles and Tortoises	Other Turtles and Tortoises	Other Turtles and Tortoises

Source: Adapted from Fink's tables II and III, except the Opossum data, which were reported by James (1959).

It is important to emphasize that *all* animals eventually succeeded on the task and, therefore, the *ability* to learn the sequence of goals in the Arrow Maze was shown to be within the capacity of all species tested. Whether the ranked orders shown represent differences in intelligence or other kinds of differences simply cannot be determined from such data. In addition to the contaminating influence of sensory, motor, and motivational differences, the animals' ages varied widely. In general, very young animals, even in relative terms, were pitted against young adults or even the relatively elderly (in the case of some of the turtles and tortoises). Once again, the tortoise (or, at least, the Eastern Painted Turtle) has outrun the hare (that is, the young New Zealand white rabbit). Was the tortoise more intelligent, or did its wisdom, experience, and perseverance overcome the impetuousness of youth?

I hope that the point has been made that direct *quantitative* differences such as the number of trials to criterion or the number of errors made while learning should not be used to compare learning abilities across species. Let us turn now to the question of possible *qualitative* differences in learning.

Qualitative Differences in Learning

Bitterman's Approach

By qualitative differences in learning, it is usually meant that some animals can perform a task successfully while others cannot, or that different animals perform the task in distinctly different ways. M. E. Bitterman and his co-workers have demonstrated, over many years, a number of qualitative differences in performance on learning tasks by a variety of vertebrates and, sometimes, invertebrates. I will illustrate this with one of his best-known examples. The example involves reversal learning of which two basic types have been used: reversal learning with spatial cues and reversal learning with visual cues.

The basic procedure in reversal learning is for the animal to be rewarded for choosing one alternative (A) rather than another (B) until a preference for A is established. Then the procedure is reversed: B rather than A is rewarded. When a preference for B has been established, A is again rewarded, and so on (Bitterman 1965). Although the different species tested required different experimental environments because of differences in sensory, motor, and motivational characteristics, Bitterman attempted to keep certain basic elements of the test apparatus analogous. With both spatial and visual cues, in each case, the animal was confronted with a pair of translucent plexiglass panels on which various colors and patterns were projected from behind, and it made its choice by pressing against one or the other of the panels in its own way: a fish might strike or bite, a pigeon peck, a monkey push with its hand, and so forth. A correct choice produced a food reward appropriate to each species.

In the spatial reversal learning task, the animal is rewarded for choosing, say, the target on the right rather than the target on the left. After it learns to respond to the target on the right, the procedure is reversed so that it must now choose the target on the left. Typically, about 20 such reversals are made. Bitterman (1965) reported that the monkeys, rats, pigeons, and turtles show progressive improvement in performance as the reversals continue, but mouth-breeder fish, cockroaches, and earthworms do not. Progressive improvement means that the animal learns to recognize in increasingly fewer trials that a reversal has occurred, and so it tends to learn each successive reversal quicker than the previous one, up to a point of optimal performance. The nonimproving animals learn each reversal as though it were a new task, that is, they take about the same number of trials to learn each reversal.

On the visual reversal task, instead of learning to respond spatially to the left or right, the animal learns to respond to one of two visual cues. For example, it might learn first to respond to a black target, then a white one, then it is reversed back to black, then back to white, and so forth, for about 20 such reversals. Note that the spatial location of the correct target must be changed randomly. This makes the visual reversal task more complicated logically compared to the spatial reversal task, because in the visual task there is one relevant cue, black or white, and also one ambiguous cue, spatial location, while in the spatial task there is one relevant cue and no ambiguous cues. Bitterman's early data showed that the monkey, rat, and pigeon showed progressive improvement on visual reversal learning, as well as spatial, but the turtle did not. His principal point was that while all the animals learned the reversals, only some showed progressive improvement that he interpreted as a qualitative difference in performance.

In Bitterman's early findings (table 2), taxonomic order appeared to be a function of the two types of tasks and whether the animal showed progressive improvement. However, as data continued to be reported it became clear that even reptiles and fish could show some progressive improvement in both types of reversal learning (Bitterman 1975), and so the early suggestion of taxonomic class difference on reversal learning disappeared. Bitterman has examined qualitative differences in a number of categories with a number of different tasks, but the qualitative differences discovered thus far do not constitute an obvious order of learning ability or intelligence.

Bitterman's approach represents the principal alternative to the traditional approach to the study of animal intelligence, which attempted to scale abilities and rank the animals in intelligence. The traditional approach failed for two reasons. One was that comparative learning abilities were too often mismeasured, as discussed earlier, and the other was that the measurement scales used were too

Table 2 Progressive Improvement in Reversal Learning

Spatial		Visual
	Yes	
Mammals		Mammals
Birds		Birds
Reptiles		
	No	
Fish		Reptiles
		Fish

Source: Adapted from Bitterman (1965).

anthropocentric, that is, too often based on conceptions about intelligence in humans. It remains to be seen whether the kinds of abilities and qualitative differences Bitterman has identified can be organized into any meaningful scale, or for that matter, whether there is any value in doing so.

Harlow's Approach

Harry Harlow, one of the best-known and most innovative students of primate behavior, attempted to measure comparative learning skills in terms of the number of ambiguous cues contained within the learning task (Harlow 1958). Harlow also referred to the ambiguous cues as error factors, which meant that these factors were potential sources of error that the animal had to eliminate in order to learn the correct solution. While Harlow's examples were visually perceived cues, theoretically other sensory modalities could be used, and the task could be administered for each particular species according to its best sensory modality.

The basic learning task in such studies is an object or cue discrimination problem in which two or more objects, shapes such as a triangle and a circle, for example, are placed over the food wells of a test tray. If the subject picked up or pushed aside the correct shape it was rewarded by finding food underneath. Figure 2 shows Harlow's example of a problem with one relevant cue (form) and one ambiguous cue (position). In any single correct trial a response is made not only to the circle, but also to the position it occupies. Because a particular position as well as a particular object is rewarded, the relevant characteristic of the chosen cue is ambiguous. During the many

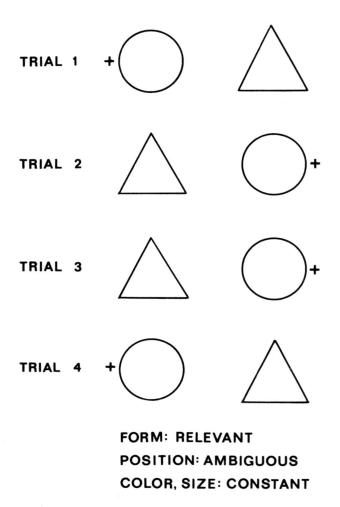

FORM: RELEVANT
POSITION: AMBIGUOUS
COLOR, SIZE: CONSTANT

Figure 2 An example of a problem with one ambiguous cue (position) that is unrelated to correctness of choice. Only the choice of the circle instead of the triangle was rewarded, whether it was on the right or the left. (Illustration by Kathleen Spagnolo.)

succeeding trials, the circle is rewarded on every trial, and each of the two positions, right and left, is rewarded on only half the trials. The inconsistent reward of the ambiguous position leads to its elimination as a reason for making a choice, and the object itself becomes the main choice. The size and color of the objects were held constant, so they were neither relevant nor ambiguous. Figure 3 shows how the problem that we just saw can be extended to two factors of ambiguity: form and position, with only the choice of color being rewarded. It should be emphasized that there is more than one way to construct a set of problems that vary in the number of ambiguous cues, and if one wanted to compare the performances of animals using ambiguous-cue problems, more than one type should be used before suggesting that an animal might be unable to perform a problem with a particular number of ambiguous cues.

Based on the data available to him, Harlow believed that all vertebrates are probably capable of successfully performing problems with one ambiguous cue. He was unsure whether any nonprimate

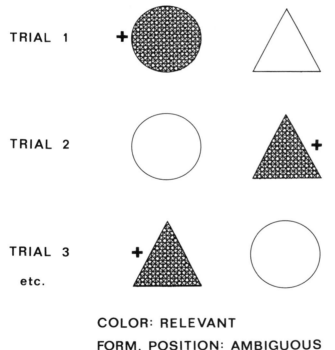

COLOR: RELEVANT

FORM, POSITION: AMBIGUOUS

SIZE: CONSTANT

Figure 3 An example of a problem with two ambiguous cues. Here the correct choice is always the darker-colored stimulus, regardless of shape or position. This is a more difficult problem than the one in figure 2 since the correct choice not only varies from right to left, but can also be either a triangle or a circle. (Illustration by Kathleen Spagnolo.)

animal had successfully performed a two-ambiguous-cue problem, but he asserted that no nonprimate animal had successfully performed a three-ambiguous-cue problem, and he noted that Henry Nissen (1951) had shown that a chimpanzee could perform a five-ambiguous-cue problem. In my laboratory, we have studied both college students and squirrel monkeys using problems similar to those used by Harlow, with two and three ambiguous cues. Both the college students and the squirrel monkeys had relative difficulty with the three ambiguous cues as compared to two ambiguous cues, but both performed significantly better than chance on both types of problems. The squirrel monkeys took about 400 trials to criterion per problem, whereas the humans took only about 34 trials on the three-ambiguous-cue problem, but of course, as noted earlier, we would not want to use this difference in the number of trials to suggest that college students are smarter than squirrel monkeys.

It should be noted that in Harlow's approach, as well as the next and last one that will be considered, the testing conditions are suited to the animal. The basic capacities to be investigated in both approaches do not depend on specific experimental tasks. Ideally, the sensory and motor requirements as well as the motivational conditions should be adapted appropriately for each species. Furthermore, the kinds of stimuli, responses, and motivating conditions used at one level should remain basically the same at succeeding levels. If the animal can perform Harlow's one-ambiguous-cue task but not his two-ambiguous-cue task, and if there is no reason to believe that the animal has deteriorated physically or that its motivation has diminished, then a failure at the higher level might reasonably be attributed to a failure in learning or a lack of intellectual capacity, rather than to sensory, motor, or motivational factors.

A Basic Hierarchy of Learning Skills

The last approach that I will consider involves a basic hierarchy of learning skills (Thomas 1980). By basic, I mean that any learning task may be reduced to or analyzed in terms of the levels of learning to be described here. If an animal is capable of learning, and if it is possible to arrange the conditions so that it will perform in the laboratory, then it should be possible to determine how far up the hierarchy of learning skills the animal is capable of performing. Since for the purposes of this paper learning is equated with intelligence, how far up the hierarchy the animal is capable of performing can be considered an index of its intelligence.

The hierarchy is essentially a synthesis of a learning hierarchy described by Robert M. Gagne (1970) and concept-learning hierarchy associated with the work of Lyle Bourne (1970) and his colleagues. Table 3 shows the result of that synthesis.

I shall pass briefly over the first four levels, because the major vertebrate classes (except amphibians, which apparently have not been tested) have been shown to be capable of performing successfully at the fifth level. Level 1, or habituation, refers to the most basic form of learning, in which an animal learns to ignore a stimulus that has no consequences. That is, it may respond at first when the stimulus is new, but as it learns that the stimulus is neither useful nor harmful, it will learn to ignore it and stop responding. This type of learning is seen in even the simplest invertebrates, as described elsewhere in this volume by Gould and Gould and by Beer. Level 2 is the same as simple classical or Pavlovian conditioning (e.g., a dog begins to salivate at the sound of a bell after the bell is repeatedly paired with food), and level 3 is simple operant conditioning (e.g., a rat presses a lever more frequently when such pressing is followed by a food reward). Level 4 refers to the chaining of simple operant responses, that is, learning more than one simple operant response in a connected sequence (e.g., a rat must press a lever and then climb up a pole before it is given a food reward). The available

Table 3   A Hierarchy of Intellective (Learning) Abilities

Relational Concepts	Level 8:	Biconditional concepts
	Level 7:	Conditional concepts
		Conjunctive concepts
		Disjunctive concepts
Class Concepts	Level 6:	Affirmative concepts
		Absolute Relative
	Level 5:	Concurrent discriminations
	Level 4:	Chaining
	Level 3:	Stimulus-Response learning
	Level 2:	Signal learning
	Level 1:	Habituation

data suggest that there are probably no fundamental differences among vertebrates in the ability to perform at the first three levels. One may expect to find differences at level 4, chaining, in terms of the length of the chains that an animal is capable of performing.

The comparison of learning ability at level 5 involves the number of concurrent discriminations an animal can learn, that is, how many simple stimulus-response discriminations it can learn and remember at the same time. For example, an animal might learn that a response to a circle is correct and a response to a square incorrect when they are both presented together. Then while still retaining that discrimination, it learns that when a triangle and oval are presented, the triangle is correct. Then a third discrimination can be added, and so on. These discriminations are considered concurrent if the animal can make the correct choice any time it is presented with any of the pairs of stimuli. Most of the data pertaining to level 5 have come from the work of Bernhard Rensch (1967) and his colleagues, who are interested in the evolution of what they call "brain achievement," which is basically the same as what I have been calling intelligence. Table 4 shows the data reported that are relevant to level 5 in the learning hierarchy. It also illustrates his belief that larger-brained species at comparable taxonomic levels will show greater evidence of brain achievement. Not shown, but interesting to note, is that the octopus, an invertebrate, has performed at level 5 with at least three concurrent discriminations.

Level 6 involves concept learning. Although there is no accepted definition of "concept," it usually refers to some common quality or characteristic shared by a number of specific stimuli that differ on one or more other characteristics. If an animal is capable of discriminating on the basis of concepts, it may be capable of practically an unlimited number of concurrent discriminations. For example, if an animal can use the concept of "tree" and the concept of "person," then it may be able to discriminate between any picture of a tree and any picture of a person. Evidence for an animal's use of a concept should be based on successful performance when new stimuli are used, or when the number of stimuli is so large that it is unlikely that the animal learned to recognize specific stimuli. A subject could learn to respond to tree A as opposed to person B, and then to tree C and not person D, and so forth, by simple concurrent discrimination learning, without ever learning the *concepts* of tree and person. But the animal that has learned the concepts can correctly rec-

Table 4 Concurrent Discrimination Learning

Animals	Number of Discriminations[a]
Teleosts (Fish)	
Perch (smaller brained)	4
Trout (larger brained)	6
Amphibians	
Not tested?	?
Reptiles	
Lizard (smaller brained)	2
Lizard (larger brained)	3
Iguana	5
Birds	
Domestic ("dwarf race")	5
Domestic ("giant race")	7
Mammals	
Mouse	7
Rat	8
Zebra	10
Donkey	13
Horse	20
Elephant	20

[a]Best as opposed to average performances.
Source: Rensch (1967).

ognize a new pair, that is, tree Y as a "tree" and person Z as a "person"; the animal that has learned only specific stimuli cannot. This example is especially appropriate, because Richard Herrnstein and his colleagues have shown that pigeons can identify pictures of specific trees and people when presented with hundreds of different slides of each (Herrnstein and Loveland 1964; Herrnstein, Loveland, and Cable 1976; see also Ristau, this volume). Since so many different slides were used, learning of specific stimuli (e.g., 1,200 concurrent discriminations!) was very unlikely.

Natural concepts, such as "trees" or "people," or concepts based on color or form involve the use of what logicians call affirmation and negation. This means that if one knows such a concept one can affirm appropriate examples of it and negate inappropriate examples ("this is a tree" or "that is not a person"). Level 6 concepts involve only affirmation and negation, and so these are called affirmative concepts or, synonymously, class concepts. However, there are two distinct kinds of affirmative or class concepts: those involving judgments based on *absolute* stimulus properties, and those involving judgments based on *relative* stimulus properties. The distinction is based on the necessity of comparing stimuli in order to make a correct choice. If you are shown a picture of a tree, you need look no further to affirm that it is a member of the class "tree." However, if you are supposed to affirm the stimulus that manifests the oddity concept (i.e., the one that is different from the rest) or the concept of "larger," you must compare your choices to determine which is the odd stimulus or which is the larger one.

To the best of my knowledge, conclusive demonstrations of the use of class concepts by animals have been limited to pigeons and primates, and pigeons have only been shown to perform absolute class concepts. Some studies in the 1920s and 1930s claimed to have demonstrated form concepts in rats, cats, and dogs, but in almost all cases the possibility of specific stimulus learning cannot be eliminated. There have also been attempts to demonstrate the oddity concept in rats, cats, and pigeons, and some investigators have claimed success. However, based on the most conservative analyses, these studies are also inconclusive on the grounds that specific stimulus learning may have occurred. In my laboratory, Linda Noble and I have been trying for about two years to develop a conclusive demonstration of oddity learning in the rat. We have tried both visual and olfactory experiments, but so far all that we see when the critical tests are administered is chance behavior.

The two highest levels in the learning hierarchy, levels 7 and 8, involve logical operations that define relationships among stimuli, and which may be described as relational concepts. Level 7 involves conjunctive, disjunctive, and conditional operations. Symbolically speaking, concepts involving A *and* B are conjunctive; concepts involving A *or* B are disjunctive. Chimpanzees and squirrel monkeys have been reported to use conjunctive and disjunctive concepts (Premack 1976; Wells and Deffenbacher 1967). Conditional concepts may be viewed symbolically in terms of the phrase, "if A, then B," with the additional requirement that either the antecedent, A, or the consequent, B, or both must be class concepts. A study by Riopelle and Copeland (1954) showed that rhesus monkeys could learn conditional concepts when the antecedent, A, was a class concept. Stephen

Kerr and I (1976) showed that squirrel monkeys could learn conditional concepts when the consequent, B, was conceptual, and Leonard Burdyn and I have recently finished a study in which both A and B were conceptual (Burdyn and Thomas 1984). I will use that study to illustrate the use of conditional concepts.

There were actually two conditional relationships in the Thomas and Burdyn study. Figure 4 illustrates the device that tested the conditional, "if septagonal, then different." On other trials, a triangle might appear in the center door, in which case the conditional was "if triangle, then same." A large number of different triangles and septagonals was used in the latter phases of the study, so the concepts "triangularity" and "septagonality" were the relevant cues, rather than specific triangles or septagonals. Similarly, the objects

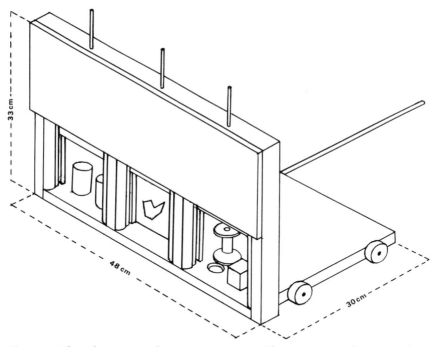

Figure 4 The Thomas-Burdyn test apparatus. The center window provides the cue indicating which of the two sets of stimuli in the other two windows is correct. The subject must move one of the objects of the correct pair to get the reward hidden beneath it. In the illustration, the spool-shaped member of the "different" pair has been displaced to show the food cup below it. The two cylinders in the other windows represent the "same" pair. (Illustration by Kathleen Spagnolo.)

that represented the concepts "same" and "different" were changed on each trial. In some stages of this study, such as that illustrated in the figure, the cue and the objects were simultaneously present, but in other stages the cue was withdrawn before the objects were presented. This meant that the squirrel monkeys had to remember both the symbol in the center door and what it stood for in order to make the correct choice. We systematically increased the time delays, and the best performance was by a monkey who could perform accurately with delays of up to 16 seconds. When we increased the delay to 32 seconds, the monkey could not learn to respond correctly in the 300 trials allowed.

Level 8 in the hierarchy of learning abilities involves the biconditional, which may be verbalized as "A if and only if B" or as "if A, then B; if B, then A." Apparently, there have been no attempts to study the ability of nonhuman animals to use biconditional concepts. I believe that it is feasible to do so, and my guess is that at least some monkeys and apes will be able to perform successfully.

Conclusions

There is little that we can confidently say at this time about the comparative intelligence of vertebrates. Despite a century of interest in such questions, there are too few data based on a common definition of intelligence or a standard scale of measurement. If it appears that only monkeys and apes were referred to as we considered the highest levels of learning, it must be remembered that, with few exceptions, only monkeys and apes have been studied at these levels. It is best to avoid debates such as the comparative intelligence of the horse versus the pig, or whether your neighbor's cat is smarter than your dog. The answers simply have not yet been found, and the only way that they will be provided unequivocally is by more well-controlled research that is based on an accepted definition of intelligence and a common framework of measurement.

Select Bibliography

Bitterman, M. E. 1965. Phyletic differences in learning. *American Psychologist* 20:396–410.

————. 1975. The comparative analysis of learning. *Science* 183:699–709.

Bourne, L. E., Jr. 1970. Knowing and using concepts. *Psychological Review* 77:546–56.

Burdyn, L. E., Jr., and R. K. Thomas. 1984. Conditional discrimination with conceptual, simultaneous, and successive cues in the squirrel monkey, *Saimiri sciureus. Journal of Comparative Psychology* 98:405–13.

Fink, H. K. 1954. *Mind and performance.* New York: Vantage Press.

Gagne, R. M. 1970. *The conditions of learning.* New York: Holt, Rinehart & Winston.

Harlow, H. F. 1958. The evolution of learning. In *Behavior and evolution,* ed. A. Roe and G. G. Simpson. New Haven: Yale University Press.

Herrnstein, R. J., and D. H. Loveland. 1964. Complex visual concept in the pigeon. *Science* 146:549–50.

Herrnstein, R. J., D. H. Loveland, and C. Cable. 1976. Natural concepts in pigeons. *Journal of Experimental Psychology: Animal Behavior Processes* 2:285–302.

Heath, P. L. 1967. Concept. *The encyclopedia of philosophy.* Vol. 2. New York: Free Press.

James, W. T. 1959. Behavior of the opposum in the Fink Arrow Maze. *Journal of Genetic Psychology* 94:199–203.

Kendler, H. H., and T. S. Kendler. 1975. From discrimination learning to cognitive development: A neobehavioristic odyssey. In *Handbook of learning and cognitive processes,* ed. W. K. Estes. *Volume 1. Introduction to concepts and issues.* New York: John Wiley & Sons.

Klüver, H. 1933. *Behavior mechanisms in monkeys.* Chicago: University of Chicago Press.

Nissen, H. W. 1951. Analysis of a complex conditional reaction in chimpanzee. *Journal of Comparative and Physiological Psychology* 44:9–16.

Premack, D. 1976. *Intelligence in ape and man.* Hillsdale, N.J.: Lawrence, Erlbaum Assoc., Inc.

Rensch, B. 1976. The evolution of brain achievements. *Evolutionary biology* 1:26–68.

Riopelle, A. J., and E. L. Copeland. 1954. Discrimination reversal to a sign. *Journal of Experimental Psychology* 48:143–45.

Rumbaugh, D. M., ed. 1977. *Language learning by a chimpanzee.* New York: Academic Press.

Thomas, R. K. 1980. Evolution of intelligence: An approach to its assessment. *Brain, Behavior and Evolution* 17:454–72.

Thomas, R. K., and R. S. Kerr. 1976. Conceptual conditional discrimination in *Saimiri sciureus. Animal Learning and Behavior* 4:333–36.

Warren, J. M. 1960. Reversal learning by paradise fish (*Macropodus opercularis*). *Journal of Comparative and Physiological Psychology* 53:376–78.

Wells, H., and K. Deffenbacher. 1967. Conjunctive and disjunctive concept learning in humans and squirrel monkeys. *Canadian Journal of Psychology* 21:301–8.

Reasoning and Language in Chimpanzees

Duane M. Rumbaugh and Sue Savage-Rumbaugh

There is great interest in the possibility that animals may share with us psychological states such as consciousness and self-awareness, as well as the ability to perceive cause and effect relationships and to base behavior upon intentions. Coupled with this great interest, however, is controversy. To deal with these topics at the human level is not without problems, so it is understandable that to attempt definition and study of them at the animal level is certain to cause disagreement. In fact, there are respectable quarters of the fields of psychology and animal behavior that would hold these mental states to be at best epiphenomena and at worst nothing but irrational projections of subjective, anthropocentric flights of fantasy that explain nothing. But considering the degree to which animals are similar to us in so many other ways, it is reasonable to suspect that they might also have certain mental states quite similar to ours, and if this is so then it is very important that we explore them. As we attempt to do so we might learn how to better understand these states that seem to control so much of our own lives.

All forms of life are, of course, unique—it is their uniqueness that defines their taxonomy. Nonetheless, no life form is without similarities to other forms, and it is generally acknowledged that similarity increases with relatedness. This frame of reference is broadly accepted for physical attributes—apes look more like humans than do monkeys, for example. It is less generally recognized that the same principle holds for psychological attributes. Just as evolution has brought about emergent physical characteristics, so it has surely brought about emergent psychological characteristics—both evolve together as they contribute to survival and reproductive success. And just as physical similarity increases with genetic relatedness, it is reasonable to believe that psychological processes become more similar with genetic relatedness. Therefore, whatever we believe about the important attributes of human cognitive processes, many of the same conclusions are probably true for other animals, and, in particular, they are likely to be true for animals most closely related to humankind. This means that they are more likely to be true for the ape than for any other animal. Whatever we ascribe to the dimensions of our individual experience—awareness, purposeful communication, perception of cause and effect relationships, the ability to formulate rules and principles from our experience, the ability to base our actions on intentions, and so on—we are most likely to find traces of these abilities in the psychological characteristics and behaviors of one of the animals that is most closely related to us, the chimpanzee (*Pan troglodytes* and *Pan paniscus*).

We have spent many years working very closely with chimpanzees as part of our long-term language research project. In the rest of this paper we will describe some of the data and observations that we have collected over the years that show evidence of some very humanlike cognitive abilities in our chimpanzee subjects.

Language Ability in Apes: The Lana Project

Our research is conducted at the Language Research Center of Georgia State University, which is operated in conjunction with the Yerkes Primate Center of Emory University. All of our operations are computer based and, for objectivity, all of our data are automatically recorded. There are three training rooms for the chimpanzees and a group room used for a variety of teaching and training procedures. Analogous facilities are available elsewhere for work with human subjects. There is also a rather spacious outdoor caging area (our animals have indoor/outside access and may withdraw from their lessons at will).

Our research goes back now about 14 years, when we began a project known as the Lana Project, which was named for our first pupil. Lana works at a console with a set of keys that she can push (see figure 1). The first keyboard that was used in 1971 had a 5 × 5

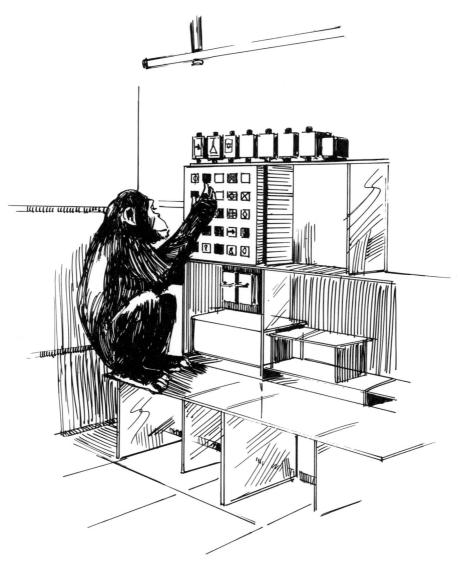

Figure 1 Lana chimpanzee at the first keyboard used in the Lana Project. Each key has a distinctive symbol that represents a word in her language. The most recent version of the keyboard has over one hundred different symbols. (Illustration by Kathleen Spagnolo, modified by Richard Swartz.)

matrix of keys; the keyboard now used has over a hundred. Each key has a distinctive symbol—a lexigram—that represents a word in Lana's language. The keys are dimly backlighted and become brighter when they are pressed. Simultaneously the keyed lexigrams are projected on overhead screens in the order pressed. By using these lexigrams, Lana can ask for various things or services—food, juice, music, even a movie. She can also ask and answer questions (Rumbaugh 1977).

Lana was trained by a method that we call "stock sentences." In other words, she was taught specific strings of lexigrams that would operate the machines and give her control over aspects of her environment. Lana gained considerable expertise in associating specific lexigrams to specific items and events, and considerable skill in mastering extensive "sentences," or strings of lexigrams. For example, one early string was "please machine give piece of banana." Anything else in the dispenser that would be vended by the piece could be requested by substituting its lexigram into the last slot of the string. With this method of substitution of symbols an attempt was made to teach her the meaning of individual symbols.

A few years ago, Thompson and Church (1980) of Brown University published a paper analyzing Lana's language ability, as reflected in the corpus of her productions published in the Lana Project book (Rumbaugh 1977). Their position was that Lana's performance could be accounted for by paired-associate learning—that is, learning to use a given lexigram with a given item or event, coupled with discriminations as to whether the experimenter was present, and if the experimenter was present, whether he was in her room or outside it but visible through a window. The overwhelming majority of Lana's productions at that time could be accommodated by that kind of explanation, which said that there were six basic stock sentences with not more than five variations on a given sentence. James Pate and one of us (DMR) (1983) recently have published a paper that takes the Thompson and Church explanation and applies it to a recent and more complex accomplishment of Lana's—one in which she was able to develop strings of her own beyond the limits of the seven-word strings of several years ago. Her system was expanded to 11-word strings, and she was given no specific training in stock sentences within that extended frame. We concluded that the Thompson and Church explanation was no longer adequate. If one takes a stock sentence approach, instead of having just six, as Thompson and Church found adequate at the time, we would have to have 69, with some of them having as many as 135 variations. The concept of the stock sentence had thus fallen under its own weight as Lana's competencies were extended.

Semantics and Syntax

Some scientists hold that language is an all-or-none phenomenon: either it is there or it is not. But research with apes has established that language should be viewed as a behavioral continuum. Competence for language builds through the independent acquisition of functions. Although there is no universally accepted working definition of language, it is generally believed that its basic requirements are *semantics,* or the use of arbitrary symbols as representations or substitutes for their actual referents in communication; and *syntax,* which refers to the ordering of symbols according to a set of rules (grammar). These two aspects of language must be considered separately in any analysis of the language ability of chimpanzees.

The contributions of ape research to semantics are rich. It is clear that the production of symbols can be at the most superficial level, devoid of any behavioral evidence that it has meaning, but we suspected that apes can do more than that. We wanted to see if our chimpanzees could learn to use lexigrams as *true symbols*—that is, if they could use them to represent things not present, a process basic to semantics in human parlance. But how might we be able to get into a chimpanzee's head to find out something about what it is thinking, if you will, as it sees lexigrams? We needed a procedure that might tell us whether or not there is a *meaning* behind these symbols, or whether, on the other hand, they are just stimuli to which we get specific, rather mechanistic responses. The experimental procedure that we developed appears to provide insight into these questions (Savage-Rumbaugh et al. 1980).

We first taught our chimpanzee subjects—who were sophisticated in language training for many years—to classify three foods and three tools, for which they had names, into two bins. A stick, money, and a key were the tools; bread, an orange, and bean cake were the foods. As these were given one by one to each of the two chimpanzees, Sherman and Austin, in turn, their task was simply to put each object into one of the two bins: one bin for foods, the other for tools. After that was learned, instead of placing items into bins, the animals were asked to press the lexigram of the item's generic definition: food versus tool (or edible versus inedible—how one translates the lexigrams is not so important as is the fact that there were two generic lexigrams for the two different categories). In addition to the foods and tools the animals had pictures of the items. After they had learned to classify these photographs accurately by pressing the correct lexigram, they were tested with photographs of other foods and tools for which they had lexigram names, and, indeed, they were able to classify them quite accurately. Sherman got nine out of nine correct on the first trial, and Austin was perfect on the second trial. They were then taught to classify the *lexigrams* of the objects used in training (see figure 2). Any time the animals were presented with a lexigram for either orange, bread, or bean cake, they were to press the key for food. And when they saw a lexigram for money, stick, or key, they were supposed to press the lexigram for tool. Again, both chimpanzees did very well at this task.

This is the limited experience of Sherman and Austin in classifying lexigrams through use of other lexigrams. What would they do, then, if they were presented with *other* lexigrams, for different foods and tools that were not even being used in the photograph phase of the study? Would they be able to look at those lexigrams and tell whether a given lexigram represented a food or a tool? They were shown the lexigrams outside of the room in which the keyboard was placed; the chimp looked at the lexigram on a given trial, went into the room on its own cognizance, and there, left only to itself, its own confidence, and a keyboard full of keys, pressed one of two keys. Both animals did very well. Austin made no mistakes whatsoever; Sherman made but one error, and that was with a sponge. We believe that one error was because the chimps often suck liquids from a sponge or chew it, and Sherman, a very vigorous eater, tends also to swallow it. There is thus a logical reason why Sherman is a bit confused about whether a sponge is a tool or a food. The chimpanzees also classified 60 items, none of which had names in their experience, as either edibles or inedibles in the same kind of paradigm, and they were usually correct—almost errorless, in fact. Overall, we feel that these results are convincing evidence that there is meaning and representation behind these symbols—at least to our chimpanzees.

It is thus clear that apes can learn to use referential symbols, a process basic to semantics in human parlance. Regarding *syntax*, it is clear that at least one ape has demonstrated great skill in ordering

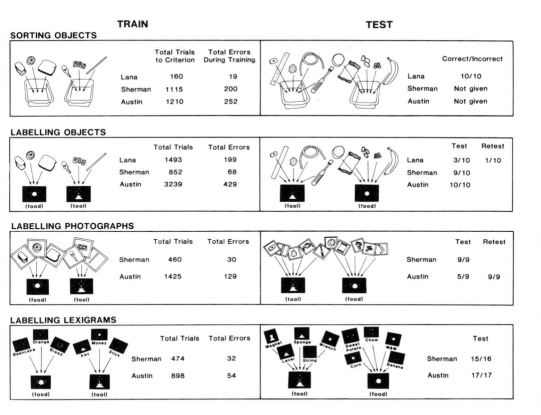

Figure 2 A diagrammatic representation of the experiment designed to determine whether chimpanzees actually understand the *meaning* of symbols or are simply responding mechanically to specific stimuli. In the first stage, the chimpanzee had to sort objects into categories; in the second, it had to press the appropriate symbol key for each object's category; and in the third stage, it had to press the appropriate category-symbol key for each photograph of an object. All of these tasks could be performed by mechanical responses to stimuli. The final stage, sorting symbols of objects into appropriate categories, requires mental representations of the objects represented by the symbols. (See text for details of the experiment.)

lexigrams (symbols) in accordance with a set of rules, which is one definition of grammar. Lana not only became proficient in ordering lexigrams in accordance with the rules, she gave clear evidence of rewriting or reordering her symbols so as to include evidence of competence for insertions and paraphrase (Stahlke 1980; Pate and Rumbaugh 1983). Admittedly, a difference remains between the facile ordering of lexigrams, or even paraphrasing through the reordering of

Reasoning and Language in Chimpanzees   63

them, and the syntax of human languages. What Lana did was somewhere well beyond the null point of having no sense of ordering of words, though somewhat short of human competence. The ability to use a grammar with competence in production and comprehension so as to encode and decode novel meaning from word order—that is, to learn how word order modulates meaning—has yet to be obtained (Pate and Rumbaugh 1983); in fact, it might not be obtainable. For the ape to be said to have some ability for language is not to say that they do have, or must have, capacity for all language functions. No chimpanzee will ever address the National Academy of Sciences.

Intentions and Inventions

What other evidence indicates that the chimpanzees know what it is that they are about? Is it reasonable to believe that they *intend* that certain things happen? The following observations suggest that they do.

There is a food-sharing situation in our laboratory in which the chimps, Sherman and Austin, are presented with a tray of delectables—10 to 15 foods, each of which has a lexigram name, placed randomly on the tray. The task has two roles, requester and vendor, and during testing the chimps exchange roles. One chimp (the requester) uses the keyboard to ask for the kind of food that he would have the other chimp (the vendor) give him. For example, the requester asks for an orange, in which case the vendor picks up two pieces of orange, gives the requester a piece, and then takes a piece for himself. Generally they keep the bigger piece for themselves—that is the advantage of being the vendor.

What happens, however, when the requester fails to ask for that which is left on the tray, as though he had forgotten the name of the item or could not readily locate it on the keyboard? This has happened several times, particularly when Austin is in the requesting role. In those instances Sherman will do more than manifest agitation. He will frequently point to the area of the keyboard where the word is. If that is not sufficient, he will take Austin by the nape of the neck and physically orient Austin's head to that portion of the keyboard. And if that is not sufficient he will take Austin's arm to guide his hand to the appropriate key so that the forefinger touches the sensor place to activate that key—and make a "request"! Such behaviors could occur, we suspect, only if Sherman has an understanding of what is necessary from Austin, based on the *intention* of obtaining a specific item on the food tray, and with the capacity to generate the appropriate repair behavior to compensate for Austin's inadequacies.

It is important to understand that Sherman and Austin have a sense of what communication can do for them, specifically communication with systems that are beyond their individual ability to create themselves. These chimps have developed no way of asking one another for a variety of foods and drinks or tools other than through the use of the lexigrams that they have been trained to use. We wondered whether they would be able to use something else for this communication if the keyboard were turned off. We conducted an experiment in which, without the keyboard, the chimps had no way of telling each other what was in a sealed container so that they might ask for it, except by using something new. The new materials were labels from food containers. We scattered the labels around on the floor. The animals had no specific training whatsoever to use food labels and no opportunity to learn what food labels were associated with, other than in the general routine of seeing foods put in or taken out of cupboards. But as soon as the keyboard was turned off, both animals, on their own cognizance, attended to the labels. They very accurately picked up a label appropriate to the item that they had seen individually put into a sealed container, and took that label over to their comrade in the adjoining room so that he, in turn, could then ask for the food by picking out from about 20 lexigrams the lexigram that corresponded to the label shown to him. Only by having some appreciation of communication beyond the means more typical of chimps, we believe, would these animals possibly have extended their efforts in the way described.

Sherman and Austin came to use their symbols as single words in much the same way as the normal child uses its first words: to clarify ambiguity and to denote the novel, or that which will communicate the most information to another (Greenfield and Savage-Rumbaugh 1984). They also will use their symbols in simple combinations in a way that cannot be satisfactorily accounted for in terms of conditioning, or satisfactorily discounted as simple single utterances appearing in close temporal proximity (Rumbaugh 1977; Stahlke 1980). We would like to give a very intriguing example of the use of a *novel combination* of words by Sherman chimpanzee that is highly suggestive of intentions and plans formulated by the chimpanzee apart from those of a structured environment. The following example is from one of Stahlke's laboratory reports:

> 2/11/83 I had just entered the colony room with my jacket on, ready to take Sherman and Austin outdoors [a very preferred activity for both of them], when Sherman, excited to see me, rushed over [to his keyboard] and asked "change TV." Janet had been showing him tapes but had *not* put in his favorite tape, of a Yerkes female, in response to "change TV." [The "change" was inserted *on the previous day* to mean change the cassette tape in the TV. I had put in this favorite tape on that day, and when I took it out Sherman had pointed to it and said "change TV."] In response to Sherman's request *today*, I changed the tape, putting in his favorite. After about 3 minutes of watching, I took the tape *out* and started to come into the room to tell Sherman I was ready to go for a walk. He gestured to me to stay where I was, which was next to the tape deck, and quickly asked again "change TV." I shook my head "no," and Janet punched into the keyboard "No change TV." And I again started into the cage. Sherman again gestured that I should stay where I was [near the tape deck] and he said "give Sherman ? change TV." In response to this I let him come *out* of the room to see what he would do. He went to the tape deck and changed the tape himself. He was very happy.

Sherman had picked up "give Sherman" from a variety of situations—he learned that it's a good thing to say "give Sherman this; give Sherman that." He had picked up "?" (question) from another training episode in which the animals asked for a variety of foods. Now they use "?food" or "?tool" to ask for things. Here Sherman put all of this together, along with "change TV," and said "Give Sherman ? change TV," which is very, very tantalizing. Somehow we have to sort out these kinds of reports and not throw them out as bathwater, lest we also lose the baby! Was Sherman asking that we give him a chance to change the TV program, and that he would rather do that than to engage in what otherwise is a highly preferred activity—that is, going outdoors for a walk? Although it cannot be concluded absolutely, it is certainly a good possibility that Sherman's repair work in this instance entailed novel use of words in a way that relates quite clearly to the behaviors subsequently manifested in that situation.

Were this the only instance of such novel use of words in combination, it might be discarded out of hand. But it is not. Many other instances have occurred in our laboratory, each of them strongly indicating integrative skills thought to be beyond those of the chimpanzee only a decade ago. These accounts demonstrate that chimpanzees not only can learn quickly and easily, but that they also have the ability to perceive or infer cause-effect relationships from past experience, and so to advantageously extend what has been learned to a relatively novel situation. This is the essence of what we mean by "invention," and we believe that ability to extrapolate what has been learned and to invent new patterns of behavior to achieve a goal are important dimensions of intelligence. Two more examples from our laboratory will further demonstrate this ability of chimpanzees to invent advantageous novel solutions.

As is characteristic of chimpanzees, Austin and Sherman have learned that when facing a mirror they are seeing reflections of themselves. Later they also learned that their images can be portrayed on a television monitor. Austin was observed when he made this discovery, and unquestionably he started to orient to the camera while also trying to maintain a good view of the monitor on which his image was projected. To do this successfully required a tradeoff between the two interests. And so, with the camera more in line with the monitor, he made sustained and well-coordinated efforts to keep his mouth open and his head oriented so that the camera focused deep into his throat, while he could see it on the monitor (figure 3). Now, a throat's ambient light level is low. Austin's solution to overcome that technical difficulty was getting a flashlight, turning it on, and shining it down into his throat! Learning of his reflection in a mirror apparently prepared him to conclude something about television cameras and monitors, and knowledge of how a flashlight works and its consequences led to its incorporation into the effort to enhance the quality of the visual image portrayed on the screen.

The next observation is an even more impressive example of invention. Sherman chimpanzee had learned, through demonstration and conditioning, to thrust a stick into a long plastic tube in order to extract a piece of orange at the far end. But one of the basic rules of the laboratory is that if a chimpanzee asks for a tool you give him that tool regardless of whether or not it is correct. One day Sherman, instead of asking for a stick when faced with the plastic tube problem, asked for a socket wrench—which looks something like a screwdriver, but has a little socket at the end for going onto the head of a

Figure 3 Austin chimpanzee uses a flashlight to illuminate his throat so as to obtain a better televised picture of it on a TV monitor. Austin "invented" a new behavior pattern in order to manipulate these tools in the way he wanted. (Illustration by Vichai Malikul, modified by Richard Swartz.)

nut. Sherman placed that wrench in the left end of the tube, only to observe the obvious: it was not long enough to push out the food at the other end. What to do? Sherman, without any trial-and-error practice, without observation of a person doing anything in that situation, extracted the wrench very carefully and put it back in, leaving the handle of the wrench protruding just a few inches from the left end. He then gave the protruding wrench a quick hit—"pow"—with his left hand, which of course served to convert that wrench into a missile that traversed the length of the tube with sufficient force to knock the food out of the other end. And, he caught the food with his right hand!

It is important to emphasize that these behaviors were not shaped by the researchers. Were they learned? Of course parts of them were; however, their integration was by the chimps and the chimps alone. If this report were of humans, no such note would be necessary. That it is of chimpanzees tells us something about considerations that must be incorporated into our theoretical perspectives of behavior.

Learning: Variations on a Theme

Early in this paper we mentioned that physical similarity among species increases with genetic relatedness, and suggested that this is true for psychological characteristics as well. Interestingly, though few people confuse the physical appearances of various life forms, some very eminent psychologists have held that laws of learning are the same across all forms of animal life. By contrast, other psychologists hold that there are important qualitative differences in learning between at least some major taxonomic groups of animals. Given the great variation in physical appearances and, more importantly, in the brains of diverse animal forms, it seems reasonable that some differences in what is learned should be expected.

For the purpose of this discussion, learning will be defined as any change of behavior due to benefits and consequences of experience that cannot otherwise be accounted for in terms of innately predisposed behaviors (e.g., tropisms, reflexes, instincts) or maturation (e.g., developmental changes in locomotion, vocalizations, dexterity). Most commonly, basic learning is believed to be due to the establishment of new stimulus-response (S-R) associations or to the establishment of new stimulus-stimulus (S-S) or response-response (R-R) associations, although the latter two are less commonly held necessary than is the first. All three of these share the principle of *association*, thus they are referred to as associative learning. Both classical and operant conditioning are examples of this type of learning.

Although associative learning is basic and important, there is also another form of learning, which we have already described for Sherman and Austin. In this type of learning, the learner is not just a recipient of stimuli, but rather a selective perceiver and organizer of information, who not only acquires associations, but also produces a cognitive structure from these associations that is capable of generating inference and new behaviors based on abstracted rules and principles. These mechanisms provide the foundation for cognition, for the "knowing about" things and the relationships between them, and they are held in this paper to be the foundation of what is termed *intelligence* in human beings.

We have recently completed a series of studies (Rumbaugh and Pate 1983) that provided evidence for the emergence of such cognitive processes during the evolution of the primate brain. We tested a wide variety of primate taxa on a complex series of object-discrimination tasks, using procedures that permitted us to assess the degree to which learning to select the correct object was accomplished by associative processes (based on the slow build-up of habit to the rewarded choice and inhibition to the nonrewarded choice) or by the learning of the rule or principle involved. Basically, the procedure was as follows.

Before the actual experiment, all subjects learned that lifting up an object in the test apparatus will sometimes uncover a reward. Then, during the *learning phase,* each subject was presented with two different objects (A and B). Choice of object A was rewarded and B not rewarded (A+B−), and we considered the task learned when the subject reached the criterion of choosing the correct object 9 times out of 10. When the criterion was reached, the *reversal phase* began: the rewarded and unrewarded objects were reversed, so that A was unrewarded and B rewarded (A−B+). The subject continued on this new task for 10 trials. Then we presented a completely new problem using a different set of objects (A' and B') and began the entire procedure again, learning to criterion followed by reversal tasks. When the second reversal was learned to criterion, a third problem was presented, and so on, for a series of 11 different pairs of objects, each with a learning phase and a reversal phase.

If the subjects had learned the original problem by simple association of A with reward and B with nonreward, reversal learning could be very difficult, since the subject must first unlearn the original associations. However, if the subject learned the basic principle involved (i.e., if one object is rewarded, continue to choose that one; if it is unrewarded, switch) then it can learn the reversal much faster, especially after a few problems. The results for the different species of primates showed that the apes were much better at these problems than the monkeys and prosimians, and performed at the same level as a group of retarded human adults who had had the benefit of language training with us. Among the chimps, Sherman and Austin, who had many years of language training in our lab, did best.

To confirm that the better performance of the apes was indeed due to the use of inference, based on the rule or principle involved rather than simply association learning, we used another set of test conditions. The original learning was the same (A+B−), but when the reward values of the objects were reversed, one of them was eliminated and replaced by a completely new object, C (either C−B+ or A−C+). If the original learning (A+B−) had occurred by the simple association of habit and inhibition with A and B, respectively, then substituting a new object for one of them should make the reversal learning easier, since the new object would be unencumbered with a "wrong" value that had been associated with it in the original learning. Alternatively, if the original learning was achieved through some rudimentary mediating or reasoning process then the three reversal conditions should be equally difficult. The subject would continue to apply the rule, and thus continue to choose a rewarded object, and switch if an object is not rewarded, whether the object is A, B, or C.

As can be seen in figure 4, there is an increase in accuracy as well as in the comparability of the three reversal conditions from monkeys to gibbons to the great apes, a progression that also reflects brain development toward the human form. These results strongly suggest that the prosimians and monkeys were using primarily associative learning, while the apes appeared able to use inferential processes. From these and other data that have been accumulating recently, including those reported in this volume by Thomas and others, there is growing evidence for the evolution of cognition associated with the evolution of the primate brain.

Concluding Remarks

This paper has addressed topics that bear upon the question: Are animals intelligent? But a prior question is: Are humans intelligent? Most will say, yes, humans *are* intelligent. Why? Because they are facile at learning, they can be clever in solving problems, they can

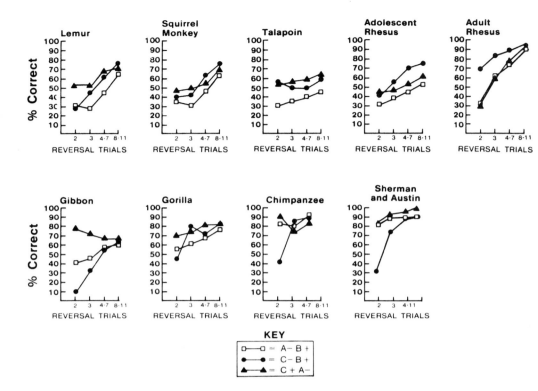

Figure 4 The relative performance of various primates on a series of discriminating learning tasks designed to determine if learning occurred as a result of the simple association of response and reward, or if some primates could learn the rule or principle involved in a complex problem and thus use inference to solve further problems. (See details of the experiment in text.)

invent new products and new ways of doing things, they are creative, they use symbols referentially and in ways to enhance communication, and so on. By the same frame of reference, it should be concluded that, yes, *animals* are intelligent to the degree that they are facile at learning, clever in solving problems, and so on.

Standards of reference, however, are always problematic, and particularly in science. Is a male bower bird clever and creative in building its nestlike bower? It is highly adaptive as it engages in its remarkable behaviors, but it should not be concluded to be intelligent in the sense of inventing bower building and extending the skills thereof to other than bower building, such as weaving fabrics for mufflers and little jackets. It is basically genetically committed to weaving only bowers. But primates show little of such genetic commitment. Certainly Austin was not genetically committed to his inventive use of the television camera, monitor, and flashlight to the end of viewing his throat. This scenario was of his own invention, based on what he apparently knew of principles and of himself and his abilities. As the object-discrimination learning experiment revealed, some species can transfer previous learning to an advantage on the basis of cognitive and at least primitive inferential processes. Intelligence per se is multifaceted and not to be explicated by any single measure, but transfer of information to a profound advantage in problem solving and invention of new patterns of behavior to effect a goal are certainly two of its main keystones.

Yes, animals are intelligent. Not so intelligent that they worry about how intelligent they are, but intelligent enough to be worthy of continued inquiry and debate.

Acknowledgments

This research is supported by National Institute of Child Health and Human Development grant HD-06016 and Animal Resources Branch of the National Institutes of Health grant RR-00165 to the Yerkes Regional Primate Research Center of Emory University.

Select Bibliography

Desmond, A. J. 1979. *The ape's reflexion.* New York: Dial Press/James Wade.

Epstein, R., R. P. Lanza, and B. F. Skinner. 1980. Symbolic communication between two pigeons (*Columba livia domestica*). *Science* 207:543–45.

Gallup, G. G., Jr. 1977. Self-recognition in primates: A comparative approach to the bidirectional properties of consciousness. *American Psychologist* 32:329–38.

Gardner, B. T., and R. A. Gardner. 1971. Two-way communication with an infant chimpanzee. In *Behavior of nonhuman primates,* ed. A. M. Schrier and F. Stollnitz. Vol. 4. New York: Academic Press.

Greenfield, P., and E. S. Savage-Rumbaugh. 1984. Perceived variability and symbol use: A common language-cognition interface in children and chimpanzees. *Journal of Comparative Psychology* 2:201–18.

Griffin, D. 1977. *The question of animal awareness.* New York: Rockefeller University Press.

Hull, C. L. 1943. *Principles of behavior.* New York: Appleton-Century-Crofts.

Mason, W. A. 1976. Environmental models and mental modes: Representational processes in the great apes and man. *American Psychologist* 31: 284–94.

Pate, J. L., and D. M. Rumbaugh. 1983. The language-like behavior of Lana chimpanzee: Is it merely discrimination and paired-associate learning? *Animal Learning and Behavior* 11(1):134–38.

Romski, M. A., R. A. White, C. E. Millen, and D. M. Rumbaugh. In press. Effects of computer-keyboard teaching on the symbolic communication of severely retarded persons: Five case studies. *Journal of Speech and Hearing Disorders.*

Rumbaugh, D. M. 1971. Evidence of qualitative differences in learning processes among primates. *Journal of Comparative and Physiological Psychology* 76:250–55.

Rumbaugh, D. M., ed. 1977. *Language learning by a chimpanzee: The LANA project.* New York: Academic Press.

Rumbaugh, D. M., and J. L. Pate. 1983. The evolution of cognition in primates: A comparative perspective. In *The evolution of primate cognition,* ed. H. L. Roitblat, T. G. Bever, and H. S. Terrace. Hillsdale, N.J.: Lawrence Erlbaum Assoc., Inc.

Savage-Rumbaugh, E. S. 1982a. Verbal behavior at a procedural level in the chimpanzee. Paper read at a meeting of the Association for Behavior Analysis, Milwaukee.

———. 1982b. A pragmatic approach to chimpanzee language studies. In *Child nurturance,* ed. H. E. Fitzgerald, J. A. Mullins, and P. Gage. Vol. 3. New York: Plenum Publishing.

Savage-Rumbaugh, E. S., J. L. Pate, J. Lawson, S. T. Smith, and S. Rosenbaum. In press. Can a chimpanzee make a statement? *Journal of Experimental Psychology: General.*

Savage-Rumbaugh, E. S., D. M. Rumbaugh, and S. Boysen. 1978a. Linguistically mediated tool use and exchange by chimpanzees (*Pan troglodytes*). *Behavioral and Brain Sciences* 1:555–57.

———. 1978b. Symbolic communication between two chimpanzees (*Pan troglodytes*). *Science* 201:641–44.

Savage-Rumbaugh, E. S., D. M. Rumbaugh, S. T. Smith, and J. Lawson. 1980. Reference: The linguistic essential. *Science* 210:922–25.

Skinner, B. F. 1953. *Science and human behavior.* New York: Macmillan.

Stahlke, H. F. W. 1980. On asking the question: Can apes learn language? In *Children's language,* ed. K. E. Nelson. Vol. 2. New York: Gardner Press.

Terrace, H. S., L. A. Petitto, R. J. Sanders, and T. G. Bever. 1979. Can an ape create a sentence? *Science* 206:891–902.

Thompson, C. R., and R. M. Church. 1980. An explanation of the language of a chimpanzee. *Science* 205:313–14.

Von Glasersfeld, E. 1977. The Yerkish language and its automatic parser. In *Language learning by a chimpanzee: The LANA project,* ed. D. M. Rumbaugh. New York: Academic Press.

The Evolution of the Brain and the Nature of Animal Intelligence

William Hodos

Many of the popular notions of animal intelligence, and indeed views held by many scientists who are inexperienced in animal behavior, can be traced to the period from the 1880s to the early 1920s. By this time the Darwinian revolution was well under way in biology and in the newly emerging field of psychology. Biologists were looking for evidence of evolutionary changes in organ systems and psychologists were interested in the behavioral heritage of humans from their animal ancestors. This was also the period during which psychologists were devising and refining the first standardized tests of human intelligence. An area of interest to both disciplines was the evolution of the brain. The notion that the brain was the organ of mind and intellect was already well established by that time, and an inescapable outcome was that early evidence about brain evolution had a great influence on early thinking about the evolution of intelligence.

Today, we use terms like "brainy" to suggest that someone has impressive intellectual power and "bird-brain" to describe the opposite. But what have we learned about brain evolution in the past 60 years? Is brain size an accurate indicator of intelligence? Do birds have unusually small brains and low intelligence? What is intelligence and how can we assess it in animals?

Early Ideas about Brain Evolution

An early pioneer of brain-evolution research was Othniel Charles Marsh, a leading paleontologist and the discoverer of many important fossils in the American Southwest (Colbert 1968). Marsh was interested in comparing the brains of fossils (as determined by impressions of the brain found on the inner surface of the cranial cavity) with the brains of living animals. He proposed several laws of brain evolution that had a major impact on his contemporaries and are still influential today (Marsh 1886). In brief, the laws state that as evolution progressed toward the present day:

—the size of the brain increased in comparison to the size of the body,

—the increase was mainly in the cerebrum (the forebrain),

—the increase in the size of the cerebrum was accompanied by a progressive increase in the number and complexity of convolutions on its surface.

Figure 1 is an illustration from a well-known textbook of human brain anatomy. It illustrates the basic points of Marsh's laws rather well: that evolution has produced brains that are relatively larger, with the greatest increases being in the cerebrum, and increased cerebral size being accompanied by increased complexity and folding of the cerebral surface. However, these brains were selected because they represent ideas of the sort that are embodied in Marsh's laws, not because they are necessarily typical of the vertebrate classes shown here. They also represent a widely held notion that evolution is a steady progression from simple to complex.

The history of brain evolution, as we understand it today, has not followed this simple scheme. Increases in brain complexity have occurred in different animal lineages and at a number of different times. They occurred in response to the specific survival needs of particular species at particular times in their histories, not as a steady evolutionary progression.

Contemporary Data on Brain Size

A contemporary approach to the study of the evolution of brain size has been to plot brain weight as a function of body weight. This serves to draw attention to relative brain size as well as absolute size (Jerison 1973) and permits scientists to determine which species have brains that are larger or smaller than would be expected for a specific body weight. Relative size would appear to be more appropriate than absolute size as a possible basis for intellectual differences

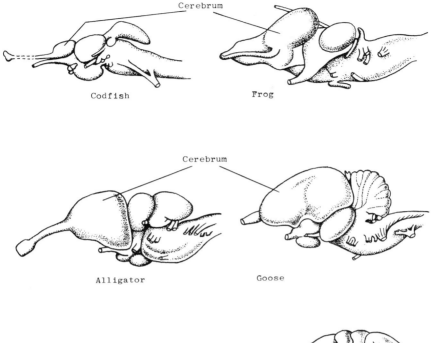

Cerebrum

Codfish

Frog

Cerebrum

Alligator

Goose

Cerebrum

Cat

Man

Figure 1 A series of vertebrate brains selected to illustrate Marsh's theory of an evolutionary progression in the relative size and complexity of the cerebrum. (Illustration by Kathleen Spagnolo, from Truex and Carpenter 1969.)

between animals because absolute size is so much affected by the size of the animal's body. A larger body would be expected to have a larger brain just as it has larger bones and larger teeth. On the other hand, if one species is more intelligent than another, its brain should be larger than we would expect it to be for an animal of that body size and weight.

This type of analysis has indicated a considerable degree of variability or diversity of brain sizes within classes and for a given body weight. Although mammals generally have the largest relative brain sizes, their brain weight distributions overlap considerably with those of birds of equivalent body weight. In other words, most birds have brains that would be regarded as typical in weight for a mammal of the same body weight. Moreover, the largest bird brains would be regarded as typical of a primate of equivalent body weight. Primates generally have the largest relative brain weights among the mammals.

Bony fishes, usually regarded without any real justification as rather lowly and stupid, have brains that are similar in weight to those of reptiles, which are regarded as much more "advanced" creatures. Indeed, the heaviest fish brains weigh more than would be expected for the heaviest reptilian brains.

Elasmobranchs (sharks and rays) also have a reputation for stupidity, but they can have surprisingly large brains: the largest elasmobranchs have brains comparable in size to those of mammals of equivalent body weight (Ebbesson 1980). But within the group there is considerable diversity. For example, although a tiger shark has more than twice the body weight of a manta ray, it has less than ⅕ the brain weight. A good question is what does the manta ray do with all of that "excess" brain?

Marsh's laws also postulate a progressive increase in the size of the cerebrum. Mammals generally do have the largest cerebrum, but again, they must share that distinction with the birds (Northcutt 1981). The cartilaginous fishes (elasmobranchs) also have cerebral weights that are quite respectable in comparison with reptiles, amphibians, and bony fishes.

But in spite of all these exceptions to the "laws" proposed by Marsh, there does appear to have been some selective pressure for larger brains, or at least for certain *parts* of the brain, in different classes of vertebrates. These parts are related to the survival needs of the animal in its particular environment. Thus, fishes have evolved rather impressive taste and smell systems to sense the chemical properties of their environments. Some have developed elaborate brain systems for the detection of the electrical fields around prey or other fishes as well as other kinds of electrical signals emitted by some fishes. Those vertebrates with superior hearing likewise have developed the appropriate brain regions for that sense. The ability to manipulate objects with the forepaws requires that substantial regions of the brain be devoted to motor control of the fingers. The high degree of sensory feedback required by this ability also necessitates a substantial commitment of brain tissues. These specializations may be only partially reflected in total brain weight or even total cerebrum weight. Figure 2 shows how the development of manual dexterity in otters is accompanied by the commitment of differing amounts of sensory cortex.

Neuroscientists often point out the importance of the cerebral cortex in various intellectual activities in humans and animals. If

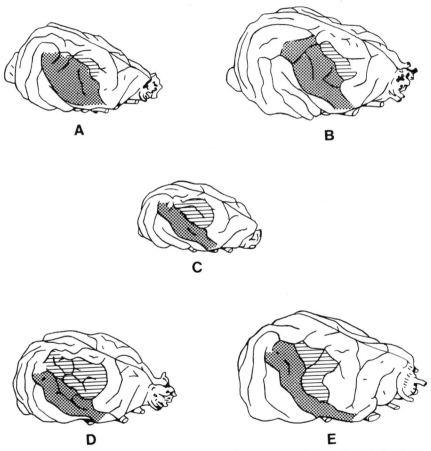

Figure 2 The sensory representation of the forepaws (striped) and the face (stippled) on the surface of the cerebral cortex of five kinds of otter (A. *Lutra*, B. *Pteroneura*, C. *Amblonyx*, D. *Anonyx*, and E. *Enhydra*). In the sequence A to E, the progressive increase in the size of the forepaw area of the cortex is correlated with an increase in the importance of the forepaws in handling food. (Illustration by Kathleen Spagnolo, from Radinsky 1968.)

one plots the volume of the cerebral cortex of various species of primates as a proportion of their total brain volume, the analysis reveals that the human cerebral cortex, about which so much has been said as an organ of intellect, occupies no greater a proportion in the human brain than in any other primate (Passingham 1982). In other words, the human cortex is about the size one would expect of a primate that large.

But what about cerebral convolutions, which figure prominently in Marsh's laws? Figure 3 shows drawings of the brains of four animals. Note that the monotreme, the rodent, and the carnivore shown have considerably more in the way of convolutions than the primate shown. To be sure, other primates have a more highly convoluted cerebrum, but so do many nonprimates, such as cows, sheep, and porpoises.

Brains do appear to be getting larger, but they are also getting more diverse, so that the range of brain sizes and adaptations is increasing. However, the benefits of these processes have not been lim-

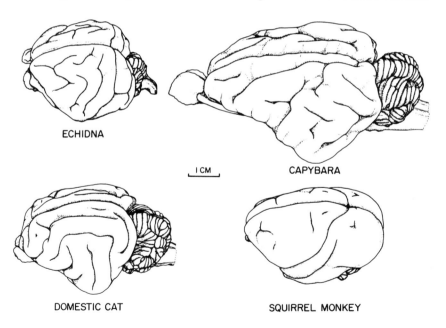

ECHIDNA

ⅼ CM

CAPYBARA

DOMESTIC CAT

SQUIRREL MONKEY

Figure 3 The brains of four mammals selected so that the primate cerebral cortex has the fewest convolutions. Such comparisons emphasize the oversimplification of Marsh's laws of progressive evolution of relative size and complexity of the cerebrum. (From Welker and Lende 1980.)

ited to primates and other mammals. Many lineages of vertebrates have developed bigger and more diverse brains (Jerison 1973).

Why should there be a selective pressure for large brains? The answer is that behavior plays an important role in adaptation to the environment. Brain components evolve in ways that permit appropriate behavioral adaptations. Animals with greater numbers of neurons arranged in more sophisticated networks probably had increased opportunities to survive and were able to pass their genes on to subsequent generations. Better sensory systems in the brain permitted responses to more subtle events in the environment. Better motor systems in the brain meant better locomotion for predation, escape, territorial defense, reproduction, parental care, and the adroit manipulation of objects. Increased integrative networks provided the capacity for improved filtering of stimuli, as well as increased storage of information and improved utilization of information so that past experience could be used more efficiently to provide new solutions to the challenges offered by the environment. In other words, successful behavioral adaptation to the environment meant brain adaptation.

Early Ideas about Animal Intelligence

About the same time that Marsh was developing his laws about brain evolution, George J. Romanes, an admirer of Darwin, published a book entitled *Animal Intelligence* (1882), which was the first attempt to deal with animal intelligence scientifically. Romanes conceptualized intelligence as an animal's capacity to adjust its behavior in accordance with changing conditions. He used phrases like "intentional adaptation" and "conscious choice" to characterize intelligent behavior. Most of Romanes's evidence was anecdotal and would not be considered as scientific by today's standards, yet his ideas about intelligence are not at variance with contemporary views. His conclusions about animal intelligence, as well as those of other writers of the period, gave rise to extravagant claims not only about the intellectual prowess of animals, but about their social, moral, and esthetic values as well. As a reaction to these excesses, C. Lloyd Morgan published *Introduction to Comparative Psychology* (1894), which was the necessary antidote to the anthropomorphizing or humanizing of animals. Morgan's idea was that one should not attribute uniquely human properties to animals if a simpler explanation would fit the facts just as well. Thus, because your dog licks your hand, you should not necessarily take this to be an expression of affection, since it may simply be an indication that the dog likes the salty taste of your skin.

A prevalent approach to human psychology at the time of Lloyd Morgan was faculty psychology, which was concerned with the mental faculties, or powers of the mind, such as perception, reason, judgment, pity, self-esteem, morality. Lloyd Morgan postulated a graded scale of these psychological attributes rising progressively from animals to humans. The influence of this psychological progression can still be felt in widely held notions about brain evolution and the evolution of intelligence.

What Is Animal Intelligence?

What is intelligence? How can we determine whether animals are intelligent or not? The first point one must consider is the distinction between "intelligence" as an abstract concept and "intelligent behavior" as something observable and measurable. The presence or absence of "intelligence" is a conclusion based on whether the observer judges the behavior to be intelligent. In other words, intelligence is not a biological property, like height or brain size; it is an abstraction based on value judgments about an organism's behavior made by an intelligence tester. If the persons or animals do not do well on the test, they are judged to have low intelligence. If they possess ample quantities of the behavioral characteristics that the observer values, they are said to be rather intelligent.

The problems of determining animal intelligence are the same as those of determining the intelligence of humans raised in cultures very different from those of the industrialized societies. Formal intelligence tests measure behavioral characteristics that we value highly for survival in our society. They do not measure characteristics that are important for survival in the indigenous cultures of New Guinea or Sudan, for example. How well would we do on tests designed by them to determine the ability to survive in their environment?

When animals are tested to determine their ability to learn a rule, the best scores are usually obtained by humans and their closest primate relatives. However, the tests are designed to be similar to tests of human intelligence, which biases them in favor of animals with good pattern vision, such as the primates. Therefore, animals such as rats, which have very poor pattern vision, do poorly on such tests. But when given the opportunity to demonstrate their rule-learning skill using odors, which they can detect far better than we can, they perform as well as rhesus monkeys do using vision. Moreover, pigeons, which have excellent vision, learn visual rules as well as primates, and birds such as mynas and blue jays are on a par with the best of the carnivores. My point is that one should judge animal intelligence not from the perspective of human behavior, but from the perspective of how well the animal is adapted to the demands of its own environment. I see little value in asking how well a crow performs as a human being.

When offering examples of animal intelligence, scientists often will point to instances of extremely complex or highly appropriate behavior, such as insect social behavior or the nest building of birds. While such illustrations are dramatic and impressive, they are not necessarily representative of intelligent behavior. They may only be manifestations of sophisticated sensory-motor programs. The hallmark of intelligent behavior, in the sense in which the term is used to describe human behavior, should be how the individual animal reacts in the face of a new challenge to its survival. Can it adapt its elaborate nest-building technique to use new materials because the old materials are no longer available? Can it develop the means to gain access to a new, but relatively inaccessible, food source when the old sources are unavailable? Can it quickly acquire new methods of predator avoidance to deal with the sudden appearance of a new form of predation that the species has never before encountered? Sometimes these new solutions will be based on past experience, that is, they will be fragments of earlier behavior patterns that have been reassembled into new behaviors. Sometimes they will be completely new patterns that appear to have no counterpart in anything that the animal has done before.

Conclusions

This discussion is not intended to suggest that animal intelligence is on a par with human intelligence. To the extent that an animal can use language or other forms of communication to enhance its survival, it has an enormous advantage over other creatures, and humans are vastly superior to all other organisms in their use of linguistic and symbolic communication. Language plays such a vital role in human intellectual activities of all sorts that any attempt to compare human intelligence to animal intelligence may be largely meaningless (Hodos 1982). On the other hand, with regard to comparisons between different classes of nonhuman vertebrates, the present evidence is such that we must seriously consider the possibility that the vast differences in intelligence that are popularly supposed to exist may in fact be relatively small (MacPhail 1982).

I would like to close by coming back to a question that I asked earlier about the legendary small brains and low intelligence of birds. Not only do birds learn rules well and have brains that would be considered quite respectable in size in a mammal, but they have been reported to make and use tools and to develop rather resourceful strategies for solving problems. Keep that in mind the next time you call someone a "bird brain." You may actually be paying them a compliment.

Acknowledgments

I am grateful to my colleague, Steven E. Brauth, for his valuable comments and suggestions. The preparation of this paper was supported, in part, by grant EY00735 from the National Eye Institute.

Select Bibliography

Armstrong, E., and D. Falk. 1982. *Primate brain evolution*. New York: Plenum.

Colbert, E. H. 1968. *Men and dinosaurs*. New York: Dutton.

Ebbesson, S. O. E. 1980. On the organization of the telencephalon in elasmobranchs. In *Comparative neurology of the telencephalon*, ed. S. O. E. Ebbesson. New York: Plenum.

Griffin, D. R. 1982. *Animal mind—Human mind*. Berlin: Springer-Verlag.

Guthrie, D. M. 1980. *Neuroethology: An introduction*. New York: Wiley.

Hodos, W. 1982. Some perspectives on the evolution of intelligence and the brain. In *Animal mind—Human mind*, ed. D. R. Griffin. Berlin: Springer-Verlag.

Jerison, H. J. 1973. *Evolution of the brain and intelligence.* New York: Academic Press.

Morgan, C. Lloyd. 1894. *Introduction to comparative psychology.* New York: Scribner.

MacPhail, E. M. 1982. *Brain and intelligence in vertebrates.* Oxford: Clarenden.

Marsh, O. C. 1886. *Dinocerata.* United States Geological Survey Monograph 10:1–243.

Northcutt, R. G. 1981. Evolution of the telencephalon in nonmammals. *Annual Review of Neuroscience* 4:301–50.

Passingham, R. 1982. *The human primate.* San Francisco: Freeman.

Radinsky, L. B. 1968. Otter brains. *Journal of Comparative Neurology* 134:495–506.

Romanes, G. J. 1882. *Animal intelligence.* London: Kegan, Paul, Trench.

Truex, R. C., and M. B. Carpenter. 1969. *Human neuroanatomy.* Baltimore: Williams & Wilkins.

Welker, W., and R. A. Lende. 1980. Thalamocortical relationships in Echidna. In *Comparative neurology of the telencephalon,* ed. S. O. E. Ebbesson. New York: Plenum.

The Sensory World of Animals

Carl Gans

A man standing in a forest at dusk perceives the temperature of the air, the shape of the surrounding trees, and the color of the setting sun. He hears insects singing and feels the humid air. A leech crawling up his foot lacks most of these senses or detects such cues but faintly, but it is thousands of times more sensitive to particular temperatures and chemical compounds, and thus it can inch steadily to the one bit of exposed skin to which it may attach. A bat flying overhead may receive very little visual information, but the bat's capacity for detection of reflected sound pulses allows it to perceive the shape of the objects in its flight path, especially the insects and other flying objects passing through this space. This spatial image may be more detailed than the visual one and is unaffected by fading light. Thus, three organisms located within a few feet of each other may perceive their world quite differently (figures 1 and 2). Those unique perceptions of the world are often referred to by the German word *Umwelt*, which can be literally translated as "surrounding world." We owe this concept to von Uexkull, who introduced the term at the turn of the century. *Umwelt* expresses the idea that any organism has only a limited concept of its environment, and that this perceptive sphere strictly reflects the information that the animal receives from its sense organs.

Senses are, of course, adaptive, by which we mean that they improve the overall survival and reproductive capacity of the organisms possessing them, and different animals have evolved different types of senses to deal with their particular environments and needs. What the zoologist aims to establish is which senses are indeed used by

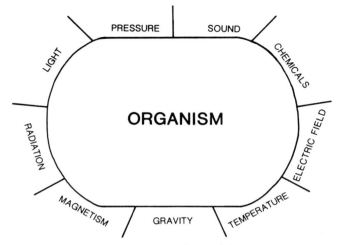

Figure 1 Elements in the perceptive sphere of living organisms. Each element, or source of stimulation, is a different form of energy to which an animal may be sensitive. How each element is perceived depends on the degree of sensitivity of one or more sense organs to that form of energy.

each kind of organism, when they are used, and what appears to be their function.

Basic Concepts about Sensory Systems

The information required by animals must answer three basic questions, although the relative importance of particular questions to particular species at any given time will vary. Question one is: What is the shape of my neighborhood? It must be answered in terms of the topography of the surrounding space and in terms of the direction and distance of assorted features from the present position. Question two is: What are the conditions of my neighborhood? A partial list of such conditions might include its moisture content, the chemical concentration and temperature of the environment, the texture of its surfaces, and the spectrum and intensity of the local electromagnetic radiation. The third question is: With whom do I share the neighborhood? Here belongs information about the detection of prey, predators, and conspecifics; not only is their existence important, but their location, size, orientation, and general nature may be critical for survival.

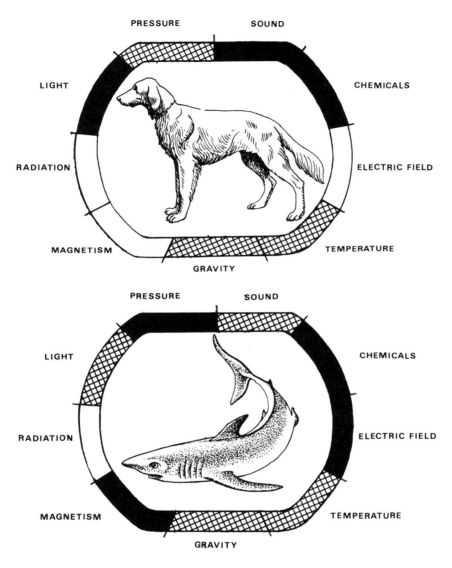

Figure 2 Differences in the degree of sensitivity of dog and shark sense organs to nine types of environmental stimuli. Black denotes great sensitivity; cross-hatching—some sensitivity; white—little or no sensitivity. (Illustration provided by C. Gans, modified by Kathleen Spagnolo.)

Sense Organs

All of this information is obtained by the use of *sense organs,* which may be defined as regions of the body that contain specialized cells and tissues that respond to incidental energy. The major sensory cells are obviously located near the surface of the animal, although animals also contain many sensory systems that monitor internal conditions. To be effective, each sensor must permit the animal to determine the *kind* of signal received (whether light is "seen" or chemicals are "smelled"), its *magnitude,* and the *direction* from which it reaches the sensor. Then this information must be communicated from the sensor to the portion of the nervous system that may integrate it and "decide" on an appropriate response.

The membranes and organelles of sensory cells respond to signals by local changes of the electrical properties of their cell surface (the membrane potential). They transmit information to the connecting nerve cells that transmit it by successive pulses along their surface membranes. Because each of these pulses depends on the fixed nature of the membrane, the output of a nerve cell cannot vary in quality (for instance, voltage of the pulse) but only in quantity: as the strength of a stimulus rises, the pulse rate either rises or falls. Impulses follow each other over a nerve membrane as waves pass over the surface of the sea; however, their height does not rise, only their spacing changes. Consequently, whether the signal a particular sensory cell generates is perceived as light or smell is determined by the way the cell is connected to the nervous system and not by any specific quality of its message.

The earliest sensory cells of animals probably were responsive to many cues; they may be thought of as having been universal detectors. Natural selection has affected the sensory system by allowing sensory cells to specialize, generating differential sensitivity to various kinds of signals. For instance, the cells within the human retina will ordinarily be triggered by light. The same cells can also be triggered by pressure—witness the flash of light perceived when a fist impacts upon the eye—but whereas a single photon of light will trigger some retinal cells, the pressure change required to produce an equivalent effect is very much higher, because the retinal cells have been selected for light reception.

Detectors face the intrinsic problem of having to detect signals varying enormously in energy. At low signal strength, the detectors have to be extremely sensitive. At high signal strength, they must keep from being overloaded. This problem has been approached in different fashions; thus, certain fishes use separate retinal cells for dim and bright light and move them into and out of position by a mechanical shift inward and outward as the light intensity changes. Other animals use selective filtration; the light-sensitive cells of their eyes contain oil droplets that serve as filters. The human cornea selectively blocks ultraviolet radiation, otherwise likely to damage retinal cells. Some frogs have special ear muscles that protect the "eardrum" when the animal is emitting high-intensity sounds that are, of course, concentrated in precisely the acoustic range in which auditory sensitivity is most advantageous.

Animals also need to determine the direction of a signal. In simplest terms, this requires only a sensor and a shield that will eliminate, block, or reduce the intensity of the incoming signals from a particular direction. One of the simplest directional detectors is seen in the light-sensitive organ of *Euglena,* a unicellular organism that has placed its single light-detecting zone adjacent to a dark, opaque spot. When the animal swims in such a way that the light hitting this zone is maximized, it approaches the light source and increases the rate of photosynthesis by its chloroplasts.

Paired receptors may allow animals to determine direction by comparison. Thus the two eyes and the two ears perceive slightly different visual and acoustic images; by comparing these they refine directionality and may even perceive depth. Such double perception is often combined with special enhancement mechanisms in the brain. The first signal to arrive blocks the reception of the second signal (perhaps from the other side or an adjacent region), thus increasing the contrast and facilitating decisions about the direction from which the signal came.

When local information is required, an animal may have multiple detectors scattered all over its body as, for instance, the pressure receptors along the lateral body line of a fish that monitor movement of nearby objects in the adjacent water. The multiplicity of receptors permits fine discrimination along the surface of the fish. In general, the more important the decisions based upon stimulation of a particular region, the greater will be the number of sense organs likely to be found there. Consequently, it is not surprising to see the concentration of many senses in the head region of early fishes and, of course, fishes today; even the lateral line system has many more canals on the head than along the body of a fish.

Once the information generated by a sense organ has been communicated to the central nervous system of the animal, it may be used in various ways. Multiple kinds of information are extracted from particular senses. We may speak of the use of signals to allow decisions that turn mechanisms on or off, or allow yes/no decisions: run or freeze? attack or veer off? For other groups of signals, it is the magnitude that is gauged in order to enhance or reduce the strength of the response. For example, in taste reception and responses to pheromones, as the strength of the chemical signal increases, so may the feeding or mating response. Then there are signals that are used as *Zeitgebers* (time setters); these allow animals to synchronize their internal clocks. Some clocks may track the daily cycle of activity and inactivity; other clocks steer seasonal cycles, reflecting the need to modify the food intake or to hibernate.

Another class of signals is used by the animal to generate maps of its surroundings. The input to the nervous system is often topographically organized—for example, an image impinges on the retina, and a somewhat distorted but geometrically similar representation of this image is transmitted (or projected) to a field on the visual cortex, called the visual projection area. Similarly, there are regions of the cerebral cortex where the touch, pain, and temperature receptors project, and these too seem roughly to retain the spatial arrangement of the skin.

A very important property of sensory maps is that they may overlie each other in the brain, for instance, along the surface of the cortex, with vertical connections (branches of neurons) connecting spatially equivalent sites. Thus, overlap has been demonstrated between the image that pit vipers see with their eyes and that perceived with the infrared receptors on their lips. A decision about the shape and color of moving objects that are radiating at temperatures of 37° C and thus differ from the environment is automatically derived.

Another important feature of sensory systems is that they often incorporate a considerable amount of redundancy. Thus, there seem to be more individual receptors than might be immediately required for a particular sense, or multiple senses may be capable of detecting a particular event. The more important an aspect of the environment is to an animal, the greater the number of sensory cells likely to be devoted to it, and the more likely is it that additional sense organs will be involved in its recognition. Some legacy of this phenomenon of redundancies remains with us; thus, one can enjoy a good wine by its color, its taste, its odor, and even its texture, and enhance this by the feel and sound of the goblet from which it is consumed.

Redundancy obviously enhances the probability of survival when some sense cannot be used or a detector has been damaged. Also, seeming redundancy may not be actual, because the several senses may be used sequentially during an approach to a place or organism. Such seeming redundancy is observed commonly when animals make important decisions about the suitability of prey or mates.

A Walk through the Senses

Let us now review some of the spectacular diversity of vertebrate sensory systems, which can be classified according to the types of energy (i.e., mechanical, electrical, chemical) to which they are sensitive.

Animals perceive both internal and external events, though in man perception of the former often bypasses the "conscious" level and is used directly to steer such important functions as heart rate, gastric motility, and perspiration.

Internal events, such as forces and movements among tissues, are detected by receptors that respond to the deformation of tissues. Simple neuron processes, various tendon and muscle spindles, and the so-called touch corpuscles are some of the receptors with this role. Other cells monitor the chemical composition of blood and urine, the pressure in blood vessels, and the temperature at various sites within the body.

Mechanical Receptors

The simplest and perhaps most primitive sensory cells of vertebrates are presumably the ciliated mechanoreceptors called neuromasts. These seem to be epithelial (surface) cells in various arrays, which bear several hairlike projections, called cilia, on their free surface, and respond when the cilia are bent. Advanced vertebrate neuromasts have directionality and mainly respond when their cilia are deflected in one direction, the reverse action having no effect or even decreasing the spontaneous rate of firing. While the cilia of some invertebrates project freely into the surrounding waters, the mechanoreceptive ciliated cells of vertebrates are generally protected in tubes or chambers. Their hairs tend to be imbedded in jelly-like substances called cupulary organs, which serve as mechanical amplifiers.

Vertebrates use such simple mechanoreceptive devices for a spectacular number of detectors. One is the lateral line organ of fishes and some amphibians that provides localized pressure reception in aquatic environments. The mechanoreceptive cells sit at the bottom of the interconnected, fluid-filled lateral line canals that open to the surface of the skin via rows of discrete pores, so that pressure events are sampled at discrete locations. We have already noted that although a single lateral line canal passes down the side of the trunk, a network of pore canals extends over the head, indicating that the animal is more "interested" in details of the environment in this region.

A second major set of mechanoreceptors involves the two sets of cavities we call the inner ears. Here bilateral comparison becomes more critical. The cilia of the several batches of mechanoreceptors are connected to a common cupula that may contain calcifications, making it denser than the surrounding tissues. The effect of gravity will load each ciliary field symmetrically or asymmetrically. Comparison of the two sides will thus provide the animal with information about its orientation with respect to the gravitational field and tell it which way is up. Naturally, short-term accelerational changes of the animal will also induce short-term pulses in the hair cell output.

Accelerational changes are also monitored by a second, more specialized portion of the inner ear, known as the semicircular canals. Almost all vertebrates have three semicircular canals on each side of the head, each located roughly at right angles to another. Both ends of each canal are closed, but one end will be slightly enlarged and contain a patch of mechanoreceptive cells called macula. The arrangement of canals in three complementary pairs then assures that acceleration of the head around any of the major axes will cause fluid to flow toward one macula and away from the matching one on the other side. This provides a much more refined record of the acceleration to which the animal exposes itself than is provided by the detectors in the large cavities.

The inner ear also contains additional detector zones that allow the animal to discern the kinds of vibrations we call sounds. Not only the source of the sound is important but also its frequency spectrum. Consequently, discriminate ability requires detection of this, as well as of the onset and cut-off of vibratory signals. The ability to detect specific signals provides the basis of communication, as cells may become narrowly tuned, primarily to detect signals of a particular pattern and frequency.

Vertebrates have developed a series of special devices that enhance the sound coming from particular directions or amplify the changes in sound pressure. The most obvious example is the eardrum, which transforms pressure changes passing down the external auditory canal into mechanical movements and then amplifies these. A chain of small bones called ossicles transmits the vibratory movements to a column of fluid. Birds and mammals suspend the sensory cells within a spiral tube (the cochlea) so that the magnitude of the signals is increased by having vibration pass in opposite directions on the two sides of the detector band. Increased tone discrimination requires more cells and consequently the band will have to be longer, so that the resulting organ might become longer than the head. This leads to a packing limitation that has been resolved by twisting the cochlea into a snail-like coil. Perhaps the most exquisite use of the hearing apparatus (violinists may disagree) is seen in the many animals, among them bats, owls, and cetaceans, that hear the reflections of emitted sound pulses and thus shape a sonar image of their sensory world.

Two specializations of the amplification system deserve mention. The first one occurs in certain burrowing lizards in which the middle ear linkage does not pass to a lateral eardrum, but rather to the lips; these animals selectively hear sounds passing toward them down a tunnel. The second specialization occurs in fish that use their air bladder for hearing, utilizing the increased mechanical deformation that occurs when air, rather than water, is compressed. Different bones, sometimes including portions of the vertebral column, form a linkage in such fish that transmits the resulting deformations to the inner ear.

Before we leave mechanoreception, we should note that various fishes, reptiles, and mammals detect movement by other methods. Fishes, some amphibians, and turtles use sensory barbels (filamentous processes) fringing the edge of the lips. Lizards and snakes detect touch by tiny horny bumps or hairs that project out of special pits in their scales. Mammals developed long facial vibrissae (snout whiskers) that are exquisitely sensitive to touch and may have been the first hairs developed by mammals.

Electrical and Magnetic Receptors

Another modification of external ciliated cells is seen in the various electroreceptors. In its simplest form, specialized cells in the skin or body covering detect the electric field changes produced by any living organisms and some chemicals. These cellular detectors, usually in multicellular arrays, respond to the electric field changes likely to reflect the presence of another organism in the vicinity. Obviously, the system must be shielded against the signals produced by the organism itself.

A variety of electroreceptive specializations appear in the teleost (bony) fishes (which include more than half of all species of vertebrates), as well as among the cartilaginous ones (sharks and rays). Many such forms multiply some of their muscle cells to emit enhanced electric pulses that establish a regular field around the fish. The field is differently deformed by conductive and nonconductive objects in the surroundings; thus, the electroreceptive detector array can also recognize the presence of prey objects that are not themselves emitting signals. We all know of the further development of this system in animals such as the electric eel that incapacitate their prey by shocking it; much more commonly, electric pulse generation is used for intraspecific communication, and the members of some fish groups recognize each other in this way.

Whereas electric field detection is well understood, our understanding of the way that some animals detect magnetic fields still poses major problems. There is good evidence that magnetic field detection occurs at least in cartilaginous fishes and salamanders, and indications that it very likely appears in birds such as pigeons. In some forms, magnetite has been identified in cells of the ear region. For others, the active principles remain unclear. Much work has been carried out on this sensory system, which, like the electric detectors, represents a category of senses apparently absent in man.

Chemical Receptors

The chemical senses of smell and taste are activated when particular end groups of molecules match the structure of the apertures in the sensing membranes. The system is exceedingly complex in terms of the cells and the kinds of compounds they recognize, and chemical signals may be perceived in a variety of places. In some teleost fishes, taste buds are distributed over much of the body surface, so that the animal may sense the spectrum of chemicals it touches. It is possible that some of the taste buds may be homologous to those found in the tongue of mammals, which detect aspects such as salty or sour. A significant fraction of other chemical information enters the central nervous system via the nasal region.

Although the several kinds of taste buds seemingly have some basic similarities, their signals may reach quite distinct portions of the central nervous system and consequently produce different behavioral responses. Various catfishes have taste buds distributed widely over their outside surface, including that covering the barbels that fringe the edges of their lips; all of these project to the facial lobe. They have a second set of taste organs in the mouth. Experimental interruption of the nerves from the external sensors keeps the fish from picking up food objects, even though it would still swallow those placed into the mouth; in contrast, interruption of the nerves from the mouth does not affect food pick up, but inhibits swallowing.

In frogs, much of the information about prey suitability appears to derive from visual cues; however, a series of taste buds lies in the upper jaw, adjacent to the internal nasal passages (or nares). These buds let frogs decide about prey suitability after the prey has reached the mouth; if the prey is noxious, the frog will eject the distasteful object.

Much olfactory information is gained through various chambers that ultimately reside along the canal from the outside of the internal nares. Several kinds of animals have two distinct systems, a vomeronasal one that mainly samples substances that are dissolved in liquids and an olfactory (smell-detecting) one that samples smaller gas-dissolved molecules. The vomeronasal system is particularly well developed in many reptiles, in which it is often involved in sex recognition. The tongue flick of snakes and lizards serves as a sampling device. Surfaces are touched by the moist tongue and the adherent particles deposited at the openings of the fluid-filled cavities in the roof of the mouth; signals pass from here to the vomeronasal sacs.

The main olfactory epithelium (scent-sensitive cells) samples most of the airborne smells. As specific cells will only detect specific compounds, it should again be obvious that a good sense of smell demands not a large nose but an extensive olfactory epithelium exposed to air changes. Indeed, this provides a morphological indicator of sensory acuity. Witness the number of small bones (the turbinals) inside the nose of dogs; the sensory surface covering them is several times larger than that in the nose of man.

Light Receptors

Perhaps the most complex of the sensory systems is the detection of radiation in the visible spectrum (meaning that part of the total that can be seen). Radiation is classified by the length of its waves. The shortest are on the electromagnetic and ultraviolent end of the spectrum and the longest on the infrared. Animals differ in their perceptive capacity; thus some birds (and bees) perceive ultraviolet, which is filtered out by the human cornea, and snakes "see" infrared, by way of pit organs that are independent of the eyes.

The simplest kind of visual organ is one that only detects the onset and cessation of illumination and integrates its intensity. This is apparently the role of the brain's pineal gland or the so-called "third eye," which apparently steers circadian and annual rhythms; we may think of the organ as telling the animal's central nervous system that spring is here or fall is approaching. Like the lateral eyes, these represent an outgrowth of the brain and are found in all vertebrates, except for a few burrowing species. Mammals have apparently retained only the central integrating portion of this system, shifting the light-detecting function to the frontal eyes.

Various kinds of eyes generally consist of an array of sensory cells arranged in a retinal surface; however, the placement of the cells is fundamentally different in vertebrates and invertebrates. In vertebrates the cells are reversed; the light-sensitive portion of each faces the opaque back of the retina, so that the light focused by the lens has to filter through several layers of connecting neurons. A process of contrast enhancement helps to compensate for the intrinsic diffusion.

This simplistic description is subject to major complications. There is, for instance, the topic of color perception or, better yet, color differentiation; that of many mammals is quite limited, with dogs and cattle perceiving some of the red colors as shades of gray. The most important role of eyes is not the detection of light but rather the detection of shape, of replicating on various parts of the brain the image that the lens forms on the retina; the most important of these projections is as a map on the visual cortex. The map will be equivalent to that projected on the retina, but reversed, thus the left input projects primarily to the right hemisphere of the brain and the right to the left. Maps have to be coordinated, specifically to develop and utilize overlap of the visual fields of the two eyes; as these see the same object from slightly different angles, the triangulation allows perception of depth.

Although it seems simple to assume that a visual map is composed of a mosaic of multicolored points that generate an image like spots on a TV screen, some elegant studies on frogs and other animals indicate that this is not the major pattern used by lower vertebrates. Instead, certain cells (or groups of cells) detect particular shapes. For instance, some selectively respond to contrasting lines of low curvature as if a large object were to move across the field. Others (appropriately called bug detectors) respond only to small objects with a highly curved perimeter. What is more interesting is that in frogs all of these signals appear to cease if the object stops moving. Does this mean that frogs may lack a spatial view of their surroundings? They must lack a detailed image. Certainly what the frog "sees" is quite different from the extremely fine visual resolution of the environment observed in many birds and mammals.

Thermal Receptors

Many vertebrates have nerve endings in their skin or body covering that respond to absolute or to changing temperature, which may in some way reflect the flow of heat between skin and environment.

Heat radiates in the infrared, and I have already mentioned that pit vipers and other snakes specifically detect this radiation by modified areas on their upper and lower lips. These sensors may be depressed into pits, and the resolution is sufficiently fine to form images equivalent to the visual ones. This permits snakes to obtain information about angle and distance of prey and thus to direct their strike. It is significant to note again that the map furnished by this system overlaps the visual one on the cortex, combining information about visual and thermal cues.

Integration

The absence of general rules, indeed, the occurrence of serendipitous usage of environmental cues, would seem to represent an important principle in the establishment of perceptive spheres. Selection appears to have been for adequacy, not for elegance or perfection. If a system worked, it was retained. Survival must often have demanded redundancy. If a message was important to the organism, more than a single sense organ is likely to have been involved in its confirmation. What we see is obviously the current stage in a process of evolution in which adaptability of perceptive capacity was important. After all, prey develop defenses and those predators that cannot overcome these will starve. Better (i.e., more perceptive) parents obviously leave more and better (i.e., more perceptive) offspring.

It may not be amiss to note that sense organs apparently have been critical in vertebrate evolution. Prevertebrate animals had an epidermal nerve network that combined sensory, integrative, and effector cells. (Effector cells activate glands or muscles.) Both the sensory and effector cells used cilia, the latter for propulsion and in filter-feeding. With the shift to muscular propulsion, these prevertebrates developed a notochord for skeletal support, becoming bilaterally symmetrical in the process. Sensory areas were concentrated anteriorly and an integrative nerve cord formed along the dorsal surface.

It has been suggested that the evolution of vertebrates resulted from a shift to active predation, leading to the development of new paired sense organs, namely the nose, eyes, ears, lateral lines, and electroreceptors. These were coupled with the gills, a new gas exchange device (lungs), and the new front end, which we recognize as the vertebrate head and which was apparently added to the front of the ancestral body by a new embryological event. This transition to vertebrates involved the invention of cartilage and bone, and it is interesting that all of these new components derive from transformed nervous tissue.

Most cells of the paired sense organs show indications in adult or embryonic stages of having developed during evolution from ciliated cells of the epithelium. Many sense organs thus provide an example for the earlier observation that selection specialized different populations of cells for particular sensory roles. The occurrence of such a common denominator in the receptors of nose, eye, ear, and even the lateral line organs, offers powerful evidence for commonality of historical origin.

So it seems that sense organs are fundamental in vertebrate organization and in vertebrate history. Apparently they have been opportunistic developments, allowing detection of environmental diversity, both physical and biological. Clearly the sense organs, and hence the aspects of the environment that animals perceive, differ profoundly among species. Consequently, information about the sensory input received by an organism is essential for understanding its mental processes.

Acknowledgments

I thank David Carrier and Peter Pridmore for their assistance with the plan for this lecture, and R. G. Northcutt for comments on the manuscript. Some of the data derive from studies supported by NSF grant DEB 8121229.

Every Animal Is the Smartest: Intelligence and the Ecological Niche

Marian Breland Bailey

Let me begin by asking a question: What do the "nest" building of a bower bird, the tail wagging of a dog, tool use by humans, and the dance of the bees have in common? Each is an example of animal behavior, more specifically of *adaptive* behavior that helps the animal survive in its environment. There are, of course, many kinds of adaptive behavior. As Gould, Beck, and others (this volume) have warned, many highly adaptive behaviors are genetically determined and cannot be considered "intelligent." But intelligent behavior, as difficult as it is to define, is almost always adaptive for an animal in its natural environment.

Such things as the ability to *discriminate*—to tell one object from another object, or to make the right response in a certain situation, which is perhaps what we mean by *judgment*—as well as the ability to remember and the ability to use symbols, all have been considered part of intelligence, and all are adaptive. But one other stands out that often determines which animal survives in a niche occupied by competitors, namely, *flexibility*. Flexibility is one of the dictionary definitions of intelligence; it is basically what psychologists mean by the ability to generalize—that is, to vary behavior in the face of changing environmental conditions, to transfer to a new set of conditions useful responses learned in other circumstances, or to see similarities between one set of stimuli and another. For example, a dolphin trained to retrieve a ring from the water's surface will readily retrieve a hat, a doll, or any similar object on the water's surface. If trained to toss a basketball with its nose, the dolphin will also use a flipper or its tail. In the long run, the flexible animal has a profound advantage when environmental changes call for a complementary rapid change in behavior.

Since 1947, our company, Animal Behavior Enterprises (ABE), has been conditioning (or training, as most people would say) animals to respond to different kinds of stimuli. Often this training has been "to order." A client might ask, "Can you train a cat to turn on a television set and play a piano?" or "Can you train a dolphin to carry packages in the open ocean?"

In the past 36 years we have been asked many questions. The two most common are "What do you think is the most intelligent animal?" and "What animal is the easiest to train?" Of course, these are almost different forms of the same question, and both are almost impossible to answer, at least as stated. To the second we must always reply, "Train to do what?" and to the first, our first answer is always, "Every animal is the smartest for the ecological niche in which it lives—if it were not, it would not be there." This is admittedly an oversimplification, but let me elaborate a bit.

Even in man's early history, animal intelligence was thought of in the context of man's own behavior—a rather anthropomorphic way of thinking. If an animal could learn tasks set for it by humans and solve human-oriented problems, it was considered intelligent; if it could not, it was stupid. But it is also common knowledge that animals are superior to humans in many abilities. Dogs, cats, rats, mice, bats, and dolphins, to name a few, hear sounds much higher in pitch than we can; bats can emit very high-pitched sounds and analyze their echoes, which helps the bat catch flying insects and avoid obstacles in flight; and dogs can sniff out the trail of prey that may be out of sight and hearing. All these characteristics influence what people think of an animal's intelligence.

There are also differences between individuals and species in the ability to associate certain stimuli with certain responses (*connectability*). These are by no means as clear to the casual observer as are the other differences I have mentioned in regard to sensory and motor abilities. However, if a dog can learn to come to its master when it hears a certain whistle, and almost always does so, that dog is considered "smart." If it cannot, it is considered "dumb." Pigs are sometimes thought "stupid" because they are "stubborn"—that is, they do not always go in the direction the farmer would like them to go. Thus, connectability certainly seems to be an important part of "intelligence," and of adaptability.

It was once believed by a large number of behavioral psychologists that any animal could learn anything, provided the correct methods were used, and that the differences between species were relatively insignificant. B. F. Skinner has pointed out that under certain laboratory conditions, all animals produce very similar patterns of responses. He writes:

> Pigeon, rat, monkey, which is which? It doesn't matter. . . . Once you have allowed for differences in the ways in which they make contact with the environment (the stimuli they respond to), and in the ways in which they act upon the environment (the responses they make), what remains of their behavior shows astonishingly similar properties (Skinner 1959).

And from another behavioral scientist:

> We arbitrarily choose almost any act from the animal's repertoire and reinforce it with food, water, or whatever else the animal will work to obtain. . . . The same act can be used for any reinforcement. . . . In any operant situation, the stimulus, the response, and the reinforcement are completely arbitrary and interchangeable. No one of them bears any biologically fixed connection to the others (Teitelbaum 1966).

Statements like these were commonly made in the 1950s and early 1960s; they resulted principally from the narrow concentration on a limited number of species—especially white rats and pigeons—and from the ardent belief of many early behaviorists that all animal and human behavior could be summarized and explained by a few basic principles.

In the late 1940s and early 1950s Keller Breland and I, like many other psychologists, confidently thought we could train any animal to do anything—although we cagily inserted in our first proposal the phrase "within the physical and neurological limits of the animal." It was well we did. For in a very few months it became clear that not all animals would condition with equal ease to react in certain ways to certain stimuli. Our paths were strewn with records of pigs that would readily pick things up but only reluctantly put them down; with cows that would *not* hurry to food, no matter how hungry they were; with chickens that would *not* stand still, when that is all they were required to do to get food (for which we replaced them with rabbits, animals that *can* sit still); with raccoons that delayed reinforcement by "washing" everything they got their paws on; and with chickens that caused the same kind of delay by shaking and cracking the plastic capsules that earned their rewards. We reported a few of these "misbehaviors" in a paper published in the *American Psychologist* (Breland and Breland 1961).

While all this was going on, we became aware of a breath of fresh air blowing across the ocean from European ethologists—zoologists who study animal behavior. The psychologist Verplanck published an article about some of the ethologists' early work on instinctive behavior patterns (Verplanck 1955). Here at last were possible answers to some of our dilemmas, and ethology has radically altered the viewpoint of many psychologists. The last twenty years or so have seen widespread acceptance of the notion that species differences *are* behaviorally significant: that animals differ in their capacities to react to certain stimuli, and that not all responses are possible for all animals. But most important is the understanding that the idea of a *general* "learning ability" is inappropriate. It is now clear that there is a qualitative difference in the types of associations the various species are capable of forming—a particular animal might make some connections readily, some with great difficulty, and some not at all. Examples have been steadily accumulating from modern psychological laboratories. Garcia and his colleagues found that with rats, food avoidance learning using taste is very fast, but food avoidance using visual or auditory stimuli is very slow (Garcia and Koelhing 1966; Garcia, McGowan, and Green 1972). On the other hand, other researchers have found that visually oriented birds quickly learned food avoidance using visual cues, but not with auditory cues (Shettleworth 1972).

All these findings from various sources have some bearing on judgments of an animal's intelligence. To be "intelligent" an animal must at least be capable of making the associations or connections that determine its survival in its niche. Simply because an animal cannot with ease make a certain kind of connection does not necessarily mean that the animal is stupid; perhaps in that animal's niche that *kind* of connection is simply not required.

Since the 1960s we at ABE have made a regular practice of studying an animal ethologically before we start a training program. We find out what types of behaviors the animal uses in its niche, what is easy to condition, what is hard. And so we trained an ungulate to use a grazing response to play a harmonica, a pig to "play a piano" by moving its feet along the keys (figure 1), a duck to paddle a boat, and a goat to "use its head" to ring a bell. For the Montreal World's Fair, Expo 67, the Canadian government asked us to show the public how various domestic animals make their living in a common ecological niche—the barnyard. The particular behaviors involved crossing a stream of water, getting food, and returning home. A pigeon flew across the stream, lighted on a small building (where it rang a bell), and flew back. A chicken walked across a log, chased a butterfly in a tree, and returned to its chicken house. A rabbit crossed the stream by going *through* the log, scratched in a garden, and returned to its

Figure 1 Piano-playing pig. The piglet uses its feet to play the piano; however, the rooting and biting responses in food-getting situations are very strong and often interfere with foot movements. (Illustration by Richard Swartz.)

burrow. And a duck slid down through a waterfall, dabbled in a pool for a fish, and then returned to its nest. All this was completely automated—controlled completely by timers and the sequence of events.

Here we need to sound a word of warning: using an animal's natural behavior in a situation like this does not necessarily demonstrate a high degree of intelligence on the part of the animal, although people generally tend to read intelligence into such a situation. Thus intelligence, in this instance, is in the eye of the beholder. We hear comments such as "Look at that smart pig playing the piano!" "I never knew a chicken was smart enough to use a computer," and so on.

For some training programs, we can also make use of the animal's ease in making certain connections. For example, a project for the U.S. Army involved training birds to fly down the road ahead of a convoy and detect camouflaged troops in ambush on the side of the road. The bird of choice here was the homing pigeon, because of its strong flight capabilities and inbred tendencies to return to the home loft (in this case a mobile loft mounted on one of the convoy's trucks). In another Army project, the assignment was to train a dog to detect buried antipersonnel mines. A hunting dog (with a keen nose) was the obvious choice; the dog was trained to sit at the location where it sniffed out a mine.

Seabirds have been scanning the ocean for fish for thousands of years. Could they also be trained to scan for other objects? Could a herring gull, for example, spot a downed flier wearing a life jacket? We decided we would find out. We started with chicks a few hours old, and raised them from infancy with human trainers. The training started in flight pens, moved to sheltered bays and small boats, and finally to the open ocean. And our hunches were right: because of the birds' strong natural hunting behavior it was easy to train them to scan for life jackets, which, through our conditioning, became associated with food.

Another aspect of behavior ABE has used to advantage is the relationship between connectability and the flexibility, or generalizing ability, which I have already mentioned. The more variable the niche, the more types of connections the animal may be required to make. We are all familiar with the countless stories of "smart" dogs locating lost masters or their belongings. We once trained a crow to retrieve a small object. On one occasion, when it could not locate the article (which was not there), it spent several extra minutes hunting, and finally returned with another article of similar size and shape.

We have, over the years, studied and trained many animals, both flexible and rigid. Some of the more inflexible are the peafowl, the swan, and the booby. The last bird is an interesting case, because in spite of relative inflexibility—as far as conditioning new behaviors—it is superbly adapted, in its flight abilities particularly, to its own niche. Some of the most flexible birds (one is tempted to say "the smartest") are the crows and their relatives, and the parrot family. In a comparative study of long-distance pattern vision, the raven proved able to master the problem readily, while the seagull, with perhaps very similar visual abilities, was a total "dunce."

One interesting contrast occurs when two different species are trained on the same type of response—a chicken, for example, makes a fine visual discrimination in picking the face card of a group of five. It is much more difficult to teach the same type of discrimination to a macaw, but the macaw can learn essentially the same behavior making a tactile discrimination (figure 2). For the chicken, the face card is marked with a tiny black dot; for the macaw, the face card bears a small indentation for which the bird feels with its tongue. Ducks, although in many ways quite different from chickens, can also learn fine visual discriminations—for example, a duck can be trained to select a picture of another duck of its own kind. This does *not* mean that the duck really sees this two-dimensional drawing as a real duck; it has simply been conditioned to make a certain type of dabbling response to a certain visual pattern.

Actually, the ability to form discriminations does not seem to

Figure 2 A macaw picks out the ace of hearts by feeling with its tongue a small indentation on the back of the card. (Illustration by Richard Swartz.)

be one of the main facets of intelligence. Many people assume that the chicken must after all be very smart because we have been able to train so many fine visual discriminations. We feel the ability to generalize, to be flexible, and to make many kinds of connections is much more "intelligent" than the capacity for forming discriminations. Among the mammals, the raccoon shows this flexibility of behavior (figure 3). This animal, like monkeys and apes, has very dextrous hands that make possible many skills. But such dexterity is not a prerequisite for flexible, "intelligent" behavior. Some of the most interesting of all the animals, to us as well as to the general public, have been the dolphins and whales: great generalizers, tre-

Figure 3 Basketball-playing raccoon. Raccoons, among mammals, are good generalizers. They exhibit a great deal of varied, flexible behavior and can use their hands (or forepaws) very much as primates do. (Illustration by Vichai Malikul.)

mendously flexible in their behavior, and extremely well adapted to their watery niche.

I do not pretend to have provided the last word on animal intelligence, but perhaps I have been able to communicate the complexity of the problem. And, perhaps the animals described here have themselves told us something about how "smart" they are. Of course, in the long run, animals have taught us more than we have taught them.

Select Bibliography

Alcock, John. 1979. *Animal behavior: An evolutionary approach.* Sunderland, Mass.: Sinauer.

Breland, K., and M. Breland. 1961. The misbehavior of organisms. *American Psychologist* 16:681–84.

———. 1966. *Animal behavior.* New York and London: Macmillan.

Garcia, J., and R. A. Koelling. 1966. Relation of cue to consequence in avoidance learning. *Psychonomic Science* 4:123–24.

Garcia, J., B. K. McGowan, and K. M. Green. 1972. Biological constraints on conditioning. In *Biological boundaries of learning,* ed. M. E. P. Seligman and J. L. Hager. New York: Appleton-Century-Crofts.

Lorenz, K. 1952. *King Solomon's ring.* New York: Crowell.

Seligman, M. E. P., and J. L. Hager, eds. *Biological boundaries of learning.* New York: Appleton-Century-Crofts.

Shettleworth, S. J. 1972. Conditioning of domestic chickens to visual and auditory stimuli: Control of drinking by visual stimuli and control of conditioned fear by sound. In *Biological boundaries of learning,* ed. M. E. P. Seligman and J. L. Hager. New York: Appleton-Century-Crofts.

Skinner, B. F. 1959. A case history in scientific method. In *Psychology—A study of a science,* ed. S. Koch. Vol. 2. New York: McGraw-Hill.

Teitelbaum, P. 1966. The use of operant methods in the assessment and control of motivational states. In *Operant behavior: Areas of research and application,* ed. W. K. Honig. New York: Appleton-Century-Crofts.

Verplanck, W. 1955. Since learned behavior is innate and vice versa, what now? *Psychological Review* 62:139–44.

The Evolution of Intelligence: Costs and Benefits

C. G. Beer

When I agreed to try to deal with the costs and benefits of intelligence I had no idea what difficulties I should have in making sense of the phrase. I had trouble both with the concept of intelligence and with the notion that it can have costs and benefits. These difficulties have become part of the subject of my paper. The view we take of the evolution of intelligence will depend upon what we take the word to mean and how we use it. People take the word to mean different things in different contexts, and different people use it differently in the same context. Consequently there can be conceptual issues as well as factual questions involved in discussion of intelligence, whether human or animal. I shall look at intelligence from four angles, which, I think, illustrate its diversity of application and manifestation: intelligence as a concrete thing; intelligence used as a metaphor; intelligence considered as learning ability; and intelligence viewed as a social phenomenon.

Intelligence as a Thing

To speak of intelligence in the manner of my title might be taken to imply the existence of something unitary, which can be measured and compared within groups, across classes, and between species. Such has often been taken to be the case, as in much of the application and interpretation of IQ testing and the consequent legislation affecting who gets the best education, who gets let into a country as an immigrant, even, in a local and quite recent instance, who is ruled fit to have children (Gould 1981).

This concept of intelligence, however, has been judged by some to be a myth. Even among the mental testers there are those who reject analyses relating results on different tests to their "loading" on a single variable, and argue instead that the data are more plausibly accommodated by their correlations with several at least partially independent variables. There are also other arguments for resisting the reification of statistical description as physical quantity or genetic substrate. One can, for instance, take the approach that Ryle (1949) took to the concept of mind, and argue that to think of intelligence as something like body weight or heartbeat is to make a "category mistake," like thinking you were being offered a reward if someone asked you to "do it for my sake." Intelligence is perhaps better represented as a dispositional concept, like malleability or shyness, its attribution being predictive of how a person will probably perform in situations requiring exercise of reason, imagination, or discrimination. At any rate, one can describe a person or a performance as intelligent without begging questions about underlying genetic, anatomical, or physiological factors. In general it is a good idea to distinguish effects from their causes.

In addition to the dubiousness of assessing intelligence quantitatively, there can be fallacy in extrapolating from estimates of its heritability within reasonably homogeneous groups of people to conclusions about the innate basis of mean differences between diverse groups of people—races, social classes, and so on (Gould 1981, pp. 156–57). Extension of such argument to comparisons between species would be a fruitless enterprise at best. Whatever we take intelligence to be when we look for its distribution in the animal kingdom, it will be closely bound up with such capacities and processes as sensory coding and perception, memory storage and recall, internal representation, flexibility of action, ability to anticipate events, and so on, all of which present qualitative diversity irreducible to representation on a single quantitative scale. We therefore need a broad and flexible concept of animal intelligence if we are to do justice to the varieties of cognition that evolution has produced.

Intelligence as Metaphor

Besides the search for evidence of intelligence in animals there has been recurrent use of cognitive language in talk about biological processes of a purely physical nature. There was a time when the goodness of fit between an organism's attributes and its way of life was presented as argument for believing that nature is literally the work of intelligence—the intelligence of God. Thus the natural theology of William Paley and the writers of the Bridgewater Treatises prepared the way for Darwin by accumulating a mass of evidence for what he construed as evolutionary adaptation. By putting natural selection in the place of God, Darwin dispensed with reliance on supernatural intelligence to account for the appearance of design in nature. Instead of being the product of foresight, planning, and reason, such appearance, from the Darwinian perspective, is the outcome of interplay between the purely material agencies of reproductive fecundity, resource limits, and inheritable variation. However, the appearance of design remains, and many instances of it are so intricate and integrated that people still resort to mentalistic metaphors when talking about it.

"Cognitive biology" has deliberately and explicitly used this approach as a heuristic strategy. Its proponents (e.g., Goodwin 1978; Boden 1980) argue that by describing evolutionary adaptation and developmental trajectory as though they were cognitively controlled, one can arrive at insights and hypotheses that would otherwise be unlikely to suggest themselves. Thus Waddington (1972) took the position that the neo-Darwinian account of adaptation as environmental sifting of random mutations left too much to chance, and so he drew analogy with cognitive models, such as the generative grammar of Chomskian linguistics, to deal with the hierarchical ordering of outcome that he viewed the process as having.

The same tack has even been taken toward complex machines, such as chess-playing computers (Boden 1970). According to Daniel Dennett (1978) the only way to deal effectively with a good computer chess opponent is to adopt an "intentional stance" toward it: treat it as though it chooses its moves according to strategies planned in the light of knowledge of the rules, comprehension of the state of play, and intention to win the game. For Dennett this is a pragmatic strategy that leaves aside the question of whether the machine can be said really to think and know and want. However, he and some others involved in "cognitive science" have been represented as holding the view that if the performance of a program passes the Turing test of being indistinguishable from the performance of a person, then the program constitutes a mind. This position, described as "strong artificial intelligence" (Searle 1980), has been severely criticized for regarding simulation as duplication, and for taking metaphorical expression literally. Whatever the need to think of a computer as a mind, a main part of the mission of cognitive science, or artificial intelligence, is to use computer programming as a way to try to understand how material substance and physical process—a brain and neural transmission—can sustain mental awareness and intelligent thought. While the enterprise has achieved success in duplicating some cognitive functions, such as memory and certain kinds of problem solving, it has yet to match human performance on a number of others, such as visual depth perception and creative synthesis. Faced with certain kinds of problems the computer can appear so stupid as to imply that its programming must be quite different from what leads a brain to solutions in similar circumstances (Boden 1979). Again intelligence would appear to be no single unitary endowment. The differences that still persist between minds and the machines designed to mimic them argue that we are still dealing in analogies and metaphors rather than duplications and identities.

Metaphorical use of the language of human mental agency also pervades much of the discussion of behavioral ecology and sociobiology. We read of "selfish genes," "deceitful communication," "cost-benefit assessment," "altruistic sacrifice," and so forth. In most contexts it is clear that such terms are merely metaphors. For example, to describe an *Ophrys* orchid as deceiving the wasps that mistake its flowers for females of their kind, and so try to copulate with them, runs little risk of making anyone think that the plant put the flowers there with the deliberate intent to deceive.

Metaphors, however, can be tricky to keep under control. Richard Dawkins (1976) has been criticized for letting his device of describing genes as though they had selfish motives and manipulative wiles lead him into arguing that all our actions are basically selfish and manipulative in their motivation (Midgley 1979). Another point is that there are contexts in which the motivational and cognitive terms might conceivably apply in more than a metaphorical sense. When we talk of animal intelligence as we have been doing in this symposium we have intended the term to apply literally, whatever the difficulties of definition. Similarly, there are times when to describe the behavior of an animal as a case of deception or altruism may carry the connotation of intentionality that is implied when such terms are used of ourselves. At any rate, this possibility should not be muted by the metaphorical application of the terms to the consequences of action, as opposed to the motives of action.

Models and metaphors play central roles in scientific thought. Their use can involve cost and benefit. The cost is that argument by analogy, being a species of fallacy, can lead thought astray. The benefit is that ". . . the implications, suggestions, and supporting values entwined with the literal use of the metaphorical expression enable us to see a new subject in a new way" (Black 1962, p. 236). However, our concern with intelligence here is with how the term might literally apply to animals and the consequent implications regarding evolution.

Intelligence as Learning Capacity

Like instinct, intelligence is anything but a precise concept. Indeed, psychologists of the past, such as Herbert Spencer, William James, and C. Lloyd Morgan, considered instinct and intelligence to be on a continuum and consequently different in degree rather than in kind. William McDougall thought an instinct had an innate motivational core experienced as a qualitatively distinct kind of emotional arousal and consequent appetite, which is served by perceptual and motor capacities, both of which can be subject to modification by learning. Even Konrad Lorenz, who argued that instinctive acts are fundamentally distinct—genetically, developmentally, and motivationally— from all other kinds of behavior, allowed that the appetitive action undertaken for the sake of their performance could reflect learning from past experience and adjustment to present circumstance.

Learning from experience and adjustment to circumstance are ingredients of intelligence shown by virtually all animals to some degree. The survival value of these capacities is self-evident where conditions are not so constant and predictable as to be safely left to a ready-made automatic mechanism, or where there is no necessity for a fully fashioned response to be forthcoming at the first time of asking because a single error would mean that there would be no second time (as in the case of a male salticid spider's courtship of his cannibalistic female). However, the possibilities for such learning and adjustment are constrained by the capacities of the animal's sensory, neural, and motor equipment. For example, Wells (1968) found that tube-dwelling polycheate worms will show habituation of their withdrawal response to sudden decrease in light intensity, such as they experience when a shadow passes over them, and that they can even be conditioned to respond to this stimulation by emerging and feeding if it is associated with presence of food. However, these learned patterns persist for only a brief period in the case of these worms, compared with such animals as arthropods, cephalopods, and vertebrates treated in similar ways. Wells explains the difference as a reflection of the limited capacity of the worms' visual sensitivity: all they can register is change in the intensity of the light falling on them, and consequently they lack the ability to distinguish different sorts of causes of what to them is the same kind of change. Since a shadow cast this morning by a waving weed could be duplicated by a worm-eating fish this afternoon, persistence of the effects of the morning's habituation could be lethal later. Similarly, the small size of an ant's central nervous system, compared with that of a rat, is reflected in differences in how these two kinds of animals learn mazes: with its limited storage capacity the ant can master only one maze at a time, and does so by assembling a sequence of response directives that it slavishly follows until they prove to be no longer reliable; the rat, with its enormously greater information capacity, can afford more flexible strategies, which are more like heuristics than algorithms, and so can switch easily from one maze to another (Schneirla 1959). The costs and benefits of evolving this or that way of behaviorally coping with the world have to be assessed in relationship to the constraints imposed and the possibilities allowed by previous anatomical and physiological endowment.

For some psychologists (e.g., Harlow 1958), intelligence is identified with "learning capability." One of the roots of this position can be traced to Herbert Spencer's view of evolution as "the continuous adjustment of internal relations to external relations" (Spencer 1855, p. 374). Indeed, Spencer saw this as a ubiquitous pattern, manifested in the development of individuals and in the progress of societies, as well as in the history of life. Thus increase in refinement of sensory capacities in evolution steadily advanced the ability of organisms to register details of the surrounding world; closer and closer matching of stimulus-response connections to the contingencies of encountered states of affairs, whether by natural selection or learned association, made finer and finer the adaptive fit between behavior and circumstance. This view of behavioral evolution as a progressive continuity running from the simplest forms of irritability to the most sophisticated forms of intellect had great appeal in Spencer's time, especially to the psychologists who fashioned what came to be known in this country as functionalism. William James noted the vagueness of the idea but praised it for its fertility; its influence persisted in the pursuit of universal principles of learning through generations of laboratory studies of rats and pigeons in mazes and Skinner boxes. However, it is a view that represents the ideal organism as a mirror of its environment, and so tends to emphasize the dependency and passivity of the organism in relation to its world.

Julian Huxley (1942) offered a different view of evolutionary progress. He argued that advanced organisms transcend their predecessors in being capable of exerting greater degrees of control over their environments and in having greater degrees of independence from their environments, both physiologically and behaviorally. Behavior shows such independence to the extent that it is free of domination by the stimuli of the moment. Nervous systems began as transmission networks connecting receptors to the means of movement. Their subsequent evolution, as sensory and motor equipment increased in complexity, has consisted of addition of intervening functions, such as the integration of multiple sensory inputs, short- and long-term storage of information acquired through experience, and coordination of commands to the mechanisms of action. In thus taking over control of the running of the body, nervous systems evolved means of holding action in abeyance, of postponing response or inhibiting it altogether, thus uncoupling the organism from coercion by circumstance and opening the way for degrees of deliberation, assessment of situations in the light of past events, and anticipation of possible outcomes by appropriate preparation. Several writers have sought the origins of intelligence in delay of response (e.g., Halstead 1947; Stenhouse 1973; Crook 1980). Hebb (1949) built an influential psychological theory on the idea.

In comparing the brains of vertebrates Hebb noted that cognitive endowment roughly correlates with the proportion of cerebral cortex consigned to functions other than sensory reception and motor control, in particular to the ratio of so-called association cortex to sensory cortex (the A/S ratio). He also pointed out that the higher the level of cognitive capability attainable by an animal the longer it takes for the animal to reach it developmentally: "We have always known that . . . a rat grows up, and develops whatever capacities an adult rat has, in three months—or a dog in six months, whereas a chimpanzee takes ten years, and a man twenty years" (Hebb 1949, p. 113). Putting these two correlations together, in the context of his theory of the physical basis of memory, Hebb argued that, unlike the sensory and motor projection areas, in which correspondence of central to peripheral relations is built in, the association cortex is unorganized to start with, and has to be structured by the impress of experience, the process taking the longer the more there is to be brought to order. In addition to furnishing the storehouse of memory, this development involves the other cognitive functions constituting intelligence: adroitness in conjecture, judgment on whether to persist in a course of action or switch to an alternative, comparison, inference, calculation, and so forth, all of which may have to be constructed via a series of stages of assimilation and accommodation, as Piaget has described for their emergence in children. The details of Hebb's story require modification in the light of more recent discovery, but his general point still stands: intelligence is bought at the expense of quick returns on reproductive investment, at least in vertebrates.

Lengthening of the period of immaturity and concomitant dependence goes with elaboration of parental care and social structure: the insects and the primates hold the championships for invertebrate and vertebrate cognitive capability; insect and primate societies also take the prizes for social sophistication. Cephalopods, cetaceans, carnivores, elephants, and birds are in the running, and for them, too, the correlation holds between cognitive and social dispositions. In the case of the insect societies, division of labor and functional specialization of castes has gone so far as to make the dependence relationship of the individual to the group analogous to that of a cell to a body. The flexibility of behavior and sophisticated forms of communication are of a remarkable order, yet their limitations are such as often to be contrasted with vertebrate achievement as lacking intelligence in the way the term applies in the looser societies of vertebrates, especially those of primates. I shall leave J. L. Gould and C. G. Gould (this volume) to argue the point, and focus my remaining remarks mainly on the mammalian manifestations of mentality.

Intelligence as a Product of Social Evolution

Among the more notable cases cited as evidence of animal intelligence in recent years have been those involving discovery of novel procedures for dealing with food, and inventive tool use. Famous examples are the potato-washing hit on by a Japanese macaque, which other members of her group learned by imitation (figure 1), and subsequently taught to succeeding generations (Kawai 1965); and the chimpanzee techniques of using prepared twigs to fish termites out of their nests and using crushed leaves as sponges to soak up drink-

Figure 1 Sweet potato washing was initiated by one young Japanese macaque and spread to other members of the troop by imitation. Over succeeding generations this behavior became more common. (Illustration by Richard Swartz.)

ing water from otherwise inaccessible pockets (Goodall 1964). In a recent discussion of "the function of intellect" Humphrey (1976) describes such cases as instances of "practical invention" in the service of "subsistence technology." However, he argues against their representing the kind of context and function in which the more advanced forms of intelligence have their raison d'être. For him, "an animal displays intelligence when he modifies his behavior on the basis of valid inference" (Humphrey 1976, p. 304), and he thinks it doubtful that the subsistence techniques were discovered in this way rather than through trial and error, accident and imitation. Similar doubts were raised by Galef (this volume). Were practical discovery in the subsistence context the main point and source of intelligence, Humphrey says, it would be paradoxical that the most intellectually gifted of animals, the great apes, enjoy undemanding subsistence conditions (figure 2). His answer to his question returns us to the link between intelligence and society. By living in social groups, he maintains, animals such as the higher primates have the means of acquiring subsistence skills through imitation and cultural transmission, and practical knowledge of the habitat and its resources through sharing of information; intelligence has evolved as a means by which such groups can be maintained—the main function of intelligence is social rather than practical.

Living in groups involves a mix of common interest and self-interest for the individuals concerned. The benefits to be had from a sheltered lengthy learning period for the young—joint participation in protection from predation, cooperation and communication in foraging, and so on—require maintenance of the social structure, and hence adaptation of the social behavior to that end. But the individuals will also be in competition with each other, both in the evolutionary sense of contending for future genetic representation, and in the more immediate and direct sense of contending for social position, reproductive opportunity, or access to food and other resources. There can be conflicts of interest between mothers and their infants over when weaning should occur, between males and females of mated pairs with regard to cuckoldry and fidelity, between the young and the old for position in the social hierarchy. Where the social organization is as open to shifts of status as it is in most primate groups, adroitness in social interaction will be favored, including judgments of another's intentions, skill in concealing and revealing one's own intentions, calculation of when to press an advantage and when to give in to pressure, awareness of relationships between

Figure 2 A chimpanzee group in its lush tropical habitat. Chimpanzees typically live in such undemanding conditions, which suggests that their great intelligence may have evolved to fulfill the need to learn the skills required in their complex social organization. (Illustration by Richard Swartz.)

others, and anticipation of manipulative maneuvers. Humphrey argues that once such a social skill begins to enhance biological fitness it will be subject to a kind of self-generated selection for improvement as each advance in manipulative subtlety and social sensitivity will set up the conditions for the next. He even envisages the possibility that this self-generative build-up of social intelligence could lead to so much time being spent in its exercise as to interfere with other activity. As evidence for this he cites Wrangham's (1975) observation that when food became scarce in the Gombe reserve the chimpanzees there curtailed their social involvement, presumably to give more time and attention to foraging.

There is much evidence of social finesse in primate societies. I pick two examples to illustrate this. My first is from Kummer (1982), who tells of clandestine sexual activity in baboon couples who take pains to prevent the suspicions of the jealous group leader from being aroused. In the case of a hamadryas pair the female meets her partner behind a rock where the leader cannot see what they are up to; but between matings she emerges to see where the leader is, and sometimes even goes to him and presents in the customary manner, before returning to her secret tryst. In the case of a gelada baboon pair given opportunity for infidelity when the female's mate could hear but not see what was going on, the male of the philandering pair withheld the loud cries he would have given had he been involved in an "honest" situation. My second example comes from Menzel's (1974) observations of captive chimpanzees living in a large outdoor enclosure within which it was possible to manipulate the situation is such a way that one individual could know where food was stashed, but not the others. When this was the situation the chimp in the know would attempt to conceal his knowledge, or try to deceive his companions by leading them in a false direction. However, the companions soon became very sharp at reading tell-tale signs given inadvertently by the deceitful one, who thus, willy-nilly, would eventually give the game away. Skill in dissembling and perspicacity in social apprehension thus drove each other to new lengths within the group.

Such observations as these make it difficult for at least some of us to believe that all social interaction below the human level is either a matter of innate tuning between signal and response, or a matter of conditioned associations of the sorts to which learning theorists have tried to relate the evidence of animal intelligence. However, someone who tries to argue for something more can find that there is a kind of double-bind to contend with: on the one hand it will be claimed that all the evidence for the use of intelligence to arrive at novel solutions is anecdotal, while on the other hand the repetition required to obtain experimental evidence gives ground for trial-and-error interpretation. Perhaps the best defense against the tough-minded behaviorist position is to argue that, in addition to being intuitively compelling, attribution of high-level cognitive capacities such as intelligence, intentionality, even some conception of others as knowing, wanting, and feeling as one does oneself, can be of heuristic value by leading to the framing of testable questions that would otherwise have little likelihood of being asked.

In harmony with this, Dennett (1983) has tried application of his "intentional systems" approach to animal cases. I have already mentioned how he has found pragmatic justification for treating a computer as though it had wishes, plans, knowledge, reasoning capability, and so forth. He has since considered the question of whether animal social interaction of the kind that I illustrated might be analyzed in terms of levels of intentionality. To illustrate this notion of level of intentionality consider my telling someone that I feel sick, and compare it with my looking pale and shivering. Both instill the belief that I feel sick, but in the first case I intend this result, and that my audience believe that to be my intention, neither of which is true in the second case. According to Dennett (following Grice 1957, 1969), only the first kind of case counts as true communication, which comprises a minimum of three orders of intentionality: communicator A intends that recipient B understand that A wants B to believe P. Do animal signals signify in this sophisticated way, or are they all to be considered as on a par with symptoms, like shivering or looking pale, which tell about an animal's state but are not intended to do so or to be understood as so intended?

A test case used by Dennett in recent discussion is the alarm-calling of Vervet monkeys (Seyfarth, Cheney, and Marler 1980). These monkeys have three different alarm-calls, one for eagles, one for leopards, and one for snakes, and each evokes a different response when the monkeys hear them: to the eagle alarm they run to ground cover; to the leopard alarm they run up into the trees; to the snake alarm they scrutinize the ground around them. Is utterance of the alarm cry merely an involuntary expression of fright caused by seeing a predator, there being a different quality of fear and hence a different expression for each kind? If so the cry should be given irrespective of whether a monkey knows itself to be within earshot of companions or not. But in fact monkeys who think themselves alone remain silent in this situation, so it would appear that alarm-calling is intended at least to get others to take avoiding action. Are the responses to the calls automatic, like those to sign stimuli as described by ethologists? If that were the case it should make no difference who gives the alarm. However, the alarm-calls given by juvenile monkeys are generally ignored, and this goes with the fact that the juveniles are less reliable than adults in their judgment of what is dangerous, often taking harmless birds, and even falling leaves, as causes for fear. Also the response that a monkey gives to an alarm-call depends upon where the monkey is as well as on the kind of call: if it is already up a tree when it hears a leopard alarm it stays where it is. This evidence that the antipredator behavior is more than blindly tuned to the signals evoking it is further supported by observations consistent with the possibility that the monkeys attend to details bearing on the caller's intentions. When recordings of the calls were played to the monkeys from concealed loudspeakers, it was found that unless the location of the monkey from whom the calls had been recorded was unknown to those hearing the playback, they would not treat it as a genuine alarm signal. A particular individual's call coming from a place other than where that individual could be perceived to be was registered as anomalous and discounted accordingly. Putting the details together, Dennett could make a plausible case for interpreting the alarm-call interactions as involving true communication at the level of third order intentionality: the signaller conveying that it wants its companions to know that there is a predator of a specific kind in the offing, so that they will take the appropriate countermeasures. There is even anecdotal evidence of their sometimes going a step further up the intentional ladder, and using an alarm-call falsely to scatter skirmishing monkeys and so break up territorial fighting that was going badly for the caller's side (Dennett 1983).

Social awareness and intelligence of this order are rare in the animal kingdom. The only challengers to primate supremacy in this regard appear to be the cetaceans, but they are still too little understood for close comparisons to be made. Elephants and pack-hunting carnivores also show considerable subtlety and complexity in their social behavior, as, to varying degrees, do most of the group-living mammals. Among birds there is much evidence of individual recognition, and social structures based on individual relationships, yet this is mixed with much behavior that appears stupid to us, given avian powers of perception, such as the blind solicitude that the host of a cuckoo's parasitism shows toward the alien chick. In reptiles, amphibians, and fishes there are only occasional glimmerings of the kinds of social finesse found in primates.

Now recall Humphrey's theory that intelligence evolved as a means of maintaining a social structure within which techniques can be learned and information shared for the exploitation of environmental resources. Since group living is common among vertebrates of all sorts, and the sharing of experience something that might be generally useful and adaptive for resource exploitation, one wonders why, if Humphrey is right, intelligence of the primate order is so rare. No doubt one factor is the degree to which an animal's means of manipulating the environment are flexible enough to be turned to different uses. If your legs have been specialized for fast running, and your mouth for grinding grasses, as they are in the ungulates, then you have much less scope for inventing new techniques or exploiting new sources of food than the primates with their relatively unspecialized equipment. However, whales are hardly more dextrous than horses, so there must be more to it than this. Another likely possibility is that the kind of evolutionary "arms race" that Humphrey envisages for intelligence requires prior arrival at a certain level of cognitive and social sophistication by other routes before it can take off, and this has happened only rarely. But given arrival at this level, is the advantage conferred by education in subsistence technology sufficient to account for why social evolution crossed the threshold into the region of intellectual self-propulsion? As Humphrey says, the subsistence demands faced by the most intellectually gifted primate societies are simple in comparison to the complexity of the social life. An alternative or perhaps complementary suggestion is that the primate kind of intelligence has its roots in reciprocal altruism (Trivers 1971).

Recriprocal altruism consists of one animal giving aid to another not closely related to it on the principle that the other will return the favor should the need arise. The altruist puts itself at risk, or expends energy that it could use for itself, which would make no evolutionary sense unless the chance were high that it will be repaid in due course. The best authenticated case, to my knowledge, is Packer's (1977) report of reciprocation between males in the troops of baboons (*Papio anubis*) that he studied in the Gombe reserve. On numerous occasions he saw a male enlist the aid of another in directing attack on a third, often one consorting with an estrous female. A typical outcome was that the male enlisting help would go off with the female, leaving the other two to finish the fight. The solicited male thus gets the blows while his friend gets the girl. However, males that collaborated in this way were most likely to get the same kind of help in return when they sought it, which they generally did from a male whose call they had previously answered. Indeed, the males tended to form lasting alliances—relationships apparently based on each partner's confidence in the other's readiness to reciprocate help with help.

Reciprocal altruism will be favored by natural selection if the fitness of individuals engaging in it is greater than that of individuals who do not. That, of course, is a tautologous statement, but there is a tertium quid that calls it into question even so. Consider an animal who solicits and accepts help from others, but never gives any in return. It will enjoy the benefits of the system but incur none of the costs. Consequently it is likely to leave more offspring than the altruists, so, if its selfishness has a genetic basis, the selfish strategy will spread through the population. However, as selfishness thus becomes more and more common, altruism will become more and more rare, and the initial advantage of selfishness will diminish through decrease in the availability of altruists to be exploited. If this continues, a point of equilibrium will be reached where the proportion of selfish to altruistic individuals is such that there is no difference of fitness between them—the population will have arrived at what John Maynard Smith (1974) christened an "evolutionary stable strategy."

This story, however, assumes the do-gooders to be quite stupid and docile. Packer's observations of the baboons suggest that reciprocal altruism goes with individual recognition, keeping score, discrimination, and the forming of alliances, all of which counter the cheat's chances of prospering. The advantage of cheating over altruism may be so great that reciprocal altruism can be maintained only in situations where such countermeasures can be exercised, namely in groups small enough for all members to know one another individually, stable and durable enough for there to be repeated occasion for the giving and getting of aid, and made up of animals having cognitive capacities sufficient for the kinds of vigilance and calculation necessary. To rule out kin selection such groups must contain genetically unrelated individuals as well. Even in such groups, however, the risks of ostracism may sometimes be worth taking for the payoff that cheating can bring. Consequently there will be inducement for increased cunning in attempts to beat the system, which will force the altruists to sharpen their wits further in retaliation. Thus we can envisage the two strategies mutually driving one another to greater and greater lengths in an "arms race" from which the highest form of intelligence is the byproduct. Such, in fact, has been the suggestion of a number of people recently pondering the question of the origins of intellect (e.g., Trivers 1971; Crook 1980). Note that the kinds of capacity entailed by reciprocal altruism and its defenses against cheating can hardly be conceived except as intentional, including as they do one animal's persuading another to believe both that it wants help and that it intends to return the favor in the future. It may well be that the reciprocation relationships have also contributed to the evolution of social feeling, involving, as they seem to, rudiments of the sentiments of trust and fellow understanding.

If, then, we focus on the forms of intelligence closest to the human, it appears that the conditions in which they most likely evolved were quite special and have come together very few times. The animals concerned must have occupied habitats in which resources were stable and abundant enough to support social systems in which parents needed to raise relatively few offspring, in whose care they could invest considerable time and effort; the lengthy period of infant dependence provided for the learning of skills necessary for successful participation in vital activities and navigation of the passage through social life in the group; the groups were of a size such that the members could know each other individually; and the social systems were such that individuals could form relationships involving both cooperation and competition calling for development of adroitness in predicting, manipulating, probing, anticipating, and adjusting to the behavior of others. The costs and benefits of the kind of intelligence emerging from these conditions are to a large extent those that have attended its evolution. For example, limiting offspring to a few carefully tended babies runs something like the risk of putting all one's eggs in one basket; if the social education depends on relationships with specific individuals, such as the mother, their loss could wreck a child's chance of becoming a socially competent adult; reciprocal altruism, supposing it to have the importance claimed for it, is a gamble exposing the "honest" player to the risk of being duped. The benefits must have outweighed such costs, or the social patterns and the intelligence to which they have given rise would not have evolved. The protection of the group must have offset the risks of having few offspring and of those few being orphaned; advance in the flexibility and efficiency in exploiting resources must have more than made up for the expenditure of time and energy in education and care of the young in preparation for their productive and social roles; and altruism toward non-kin must have been reciprocated sufficiently more often than not for it to exist where it does.

However, unless the concepts of cost and benefit are more operationally defined in the context of a specific kind of study, such as an investigation of optimum foraging strategies, they can be as slippery as soap. At least that is what I have come to think in writing this paper. Judgment of benefit especially, it seems to me, depends largely on one's perspective. What is today's advantage can be tomorrow's liability, as the dinosaurs found to their cost. Unlike the benevolent deity envisaged by the natural theologians of the eighteenth and nineteenth centuries, nature does not look beyond the immediate odds when placing her bets. Human intelligence, which we believe to have been evolved from primate intelligence, has made us the dominant creatures on the globe and brought us all sorts of benefits, from airplanes to zoological societies. But it has also brought environmental wreckage, degrees of suffering that no other kind of animal has had to endure, and the possibility that we shall blow ourselves and the rest of nature to smithereens. The balance sheet of the costs and benefits of the evolution of intelligence has yet to be tallied.

Select Bibliography

Black, M. 1962. *Models and metaphors.* Ithaca, N.Y.: Cornell University Press.

Boden, M. A. 1970. Intentionality and physical systems. *Philosophy of Science* 37:200–14.

―――. 1979. *Artificial intelligence and natural man.* New York: Basic Books.

―――. 1980. The case for a cognitive biology. *Aristotelian Society Supplement* 54:25–49.

Crook, J. H. 1980. *The evolution of human consciousness.* Oxford: Clarendon Press.

Dawkins, R. 1976. *The selfish gene.* Oxford: Oxford University Press.

Dennett, D. 1978. *Brain storms.* Montgomery, Vt.: Bradford Books.

―――. 1983. Intentional systems in cognitive ethology: The "Panglossian Paradigm" defended. *Behavioral and Brain Sciences* 6:343–90.

Goodall, J. 1964. Tool using and aimed throwing in a community of free-living chimpanzees. *Nature* 201:1264–66.

Goodwin, B. C. 1978. A cognitive view of biological process. *Journal of Social and Biological Structures* 1:119.

Gould, S. J. 1981. *The mismeasure of man.* New York: Norton.

Grice, H. P. 1957. Meaning. *Philosophical Review* 66:377–88.

———. 1969. Utterer's meaning and intentions. *Philosophical Review* 78: 147–77.

Halstead, W. C. 1947. *Brain and intelligence.* Chicago: University of Chicago Press.

Harlow, H. F. 1958. The evolution of learning. In *Behavior and evolution,* ed. A. Roe and G. G. Simpson. New Haven: Yale University Press.

Hebb, D. O. 1949. *Organization of behavior.* New York: Wiley.

Humphrey, N. K. 1976. The social function of intellect. In *Growing points in ethology,* ed. P. P. G. Bateson and R. A. Hinde. Cambridge: Cambridge University Press.

Huxley, J. S. 1942. *Evolution: The modern synthesis.* London: Allen and Unwin.

Kawai, M. 1965. Newly acquired pre-cultural behavior of the natural troop of Japanese monkeys on Koshima Islet. *Primates* 6:1–30.

Kummer, H. 1982. Social knowledge in free-ranging primates. In *Animal mind—Human mind,* ed. D. R. Griffin. Berlin: Springer-Verlag.

Maynard Smith, J. 1974. The theory of games and the evolution of animal conflict. *Journal of Theoretical Biology* 47:209–21.

Menzel, E. W. 1974. A group of young chimpanzees in a one-acre field. In *Behavior of nonhuman primates,* ed. A. M. Schrier and F. Stollnitz. New York: Academic Press.

Midgley, M. 1979. Gene-juggling. *Philosophy* 54:439–58.

Packer, C. 1977. Reciprocal altruism in *Papio anubis. Nature* 265:441–43.

Ryle, G. 1949. *The concept of mind.* London: Hutchinson.

Schneirla, T. C. 1959. L'apprentissage et la question du conflict chez la fourmi. Comparison avec le rat. *Journal de Psychologie* 57:11–44.

Searle, J. R. 1980. Minds, brains and programs. *Behavioral and brain sciences* 3:417–24.

Seyfarth, R. M., D. L. Cheney, and P. Marler. 1980. Monkey responses to three different alarm calls: Evidence for predator classification and semantic communication in a free-ranging primate. *Animal Behavior* 28:1070–94.

Spencer, H. 1855. *The principles of psychology.* London: Longmans.

Stenhouse, D. 1973. *The evolution of intelligence.* London: Allen and Unwin.

Trivers, R. L. 1971. The evolution of reciprocal altruism. *Quarterly Review of Biology* 46:35–57.

Waddington, C. H. 1972. Epilogue. In *Toward a theoretical biology* Vol. 4, ed. C. H. Waddington. Edinburgh: Edinburgh University Press.

Wells, M. J. 1968. Sensitization and the evolution of associative learning. In *Neurobiology of invertebrates,* ed. J. Salanki. New York: Plenum Press.

Wrangham, R. W. 1975. The behavioral ecology of chimpanzees in Gombe National Park. Ph.D. diss., University of Cambridge.

Tools and Intelligence

Benjamin B. Beck

Wild chimpanzees use tools to fish termites from their subterranean nests. They insert twigs and other elongated objects into the nest openings and wait for soldiers and workers to swarm onto the probe. They then withdraw the probe and eat the attached insects (figure 1). There is another method of extracting termites from their nests. Locate a nest with a breach that is being repaired by colony members. Glue fragments of the nest material onto yourself so that you can approach the hole without alarming the insects and causing them to retreat inside. Snatch a termite worker, dangle it enticingly near the hole, and then capture and eat worker after worker as they approach to investigate and rescue the bait. The use of camouflage and bait are unusual forms of animal tool use, although they are widely used by human hunters. They are strategic and clever ways of reducing the distance between hunter and hunted and they necessitate considerable knowledge of the behavior of the prey.

I have misled you by implying that it was chimpanzees that used the camouflage and bait technique to capture termites. Actually, the hunter was another insect, a Neotropical assassin bug. The bug glues the camouflaging nest material onto the entire dorsal surface of its body, captures and dangles the worker "bait" with its forelegs, and then seizes and eats the deceived insects. This behavior has only recently been described by Elizabeth McMahan of the University of North Carolina (McMahan 1983). I misled you for a reason. Camouflage and baiting performed by a large-brained ape with clear morphological and behavioral affinities to humans would clearly indicate intelligence. In contrast, discovering that the hunter is a brainless insect somehow weakens the inference of intelligence. The choice is

Figure 1 Chimpanzees "fish" for termites by inserting a carefully selected twig or grass stalk into a hole they have uncovered in a termite mound. The defending termites attack the intruding tool and are extricated and eaten by the chimpanzees. This use of tools involves complex cognitive processes such as mental imagery, premeditation, and strategy. (Illustration by Richard Swartz.)

simple: either we accept the notion that insects can behave as intelligently as apes (or humans), or we abandon our intuitive association between tool use and intelligence. I believe that we must do a bit of both.

Conventional wisdom dictates that the manufacture and use of tools are uniquely associated with higher-order intelligence, and that behaviors not involving tools are cognitively more simple. However, the behavior of the assassin bug, which most certainly uses the nest material as a camouflaging tool and the insect bait as a tool to lure others, must be assumed to be cognitively simple. The bug has a simple nervous system. Little is known of the acquisition of the behavior, but it appears not to be learned. Indeed, the behavior has been observed to be performed only during preadult nymphal stages and the technique seems to be highly stereotyped. These attributes suggest that the behavior is innate or genetically prewired. Given a healthy and hungry assassin bug in the vicinity of a termite nest under repair, the behavior will occur. The nymph need not have had a previous opportunity to acquire any specific information through learning. Many patterns of tool use by invertebrates, as well as some by vertebrates, appear to be comparably innate and, like self-guided missiles, cannot be said to be intelligent despite their being effective (adaptive) and intricate.

Trial-and-Error Learning

Many tool-use patterns of birds and mammals are not genetically determined but rather are learned. However, most of these are also cognitively simple since, as the late K. R. L. Hall (1963) pointed out, they are based on trial-and-error learning. An example, drawn from my own research, involves the acquisition of tool use by a captive hamadryas baboon. I placed a pan of choice foods out of reach on a table in front of the baboon group's cage. I also gave them an L-shaped rod that was long enough to reach the food. The adults of the group looked at the pan, reached once or twice for it, and then simply sat back and waited to be fed as usual. The adults largely ignored the tool. The younger group members also looked and reached, but then began to play with the rod, as they would with any novel object. After about 11 hours, a young male, while playing with the rod, accidentally flipped it over the pan. When he retrieved the rod, the pan was brought within reach. Within 25 trials he was getting the food in less than 2 minutes, and by the 60th trial his solution time was about 30 seconds (figure 2).

This was clearly tool use, and no training or shaping was used to produce the discovery. Yet the underlying learning process is exactly that of a rat learning to press a bar in a Skinner box, or a dog learning to come when called. First, a successful or correct response is emitted fortuitously—in play, out of frustration, or unintentionally in the

Figure 2 A young male hamadryas baboon uses an L-shaped rod as a tool to pull a tray of food within reach. This type of tool use is cognitively simple, as it is the result of trial-and-error learning. (Illustration by Richard Swartz.)

course of other behaviors. When the response is followed closely by a reward, it is likely to be repeated. Such trial-and-error or "operant" learning is common in both invertebrates and vertebrates. It is certainly adaptive, and long and intricate chains of operant responses can be crafted by skilled trainers. Operant learning even subserves much human behavior. Yet its ubiquity prevents its use for distinguishing intelligent from unintelligent behavior, and the many forms of tool use based on operant learning must therefore be designated as cognitively simple.

I wish to emphasize that although innate and operantly learned behavior are cognitively simple, they are most certainly adaptive. Indeed, information subserving truly essential behavior that occurs in a largely invariant context (such as maternal behavior in mammals) appears to be genetically programmed. Learning is reserved for the acquisition of less critical information, or information that changes too rapidly to be absorbed genetically over generations. Intelligence seems to characterize the learning of generalists: species that occupy a broad ecological and social niche and thrive on environmental variability as opposed to stability. Intelligence is adaptive—that is, it enhances fitness—only in physical and social environments that change rapidly. In our search for a relationship between tools and intelligence, we should keep in mind that intelligence is simply an adaptive strategy, and not a superior or elevated capacity.

Recall the conventional wisdom: tool behavior is uniquely associated with cognitive complexity while behavior not involving tools is cognitively simple. That we have provided examples of tool behavior that are cognitively simple should raise some doubt about the conventional view. Further, there are many examples of behaviors that are cognitively complex but which do not involve tools. Social carnivores such as wolves, lions, and hyenas, for example, cooperatively hunt large prey, working in relays to exhaust the prey or drive it toward fellow group members hidden in ambush. Such behavior implies premeditation, intentionality, and communication about events that are removed in space and time from the present. As another example, gulls in many parts of the world drop shells on hard surfaces to break them and gain access to the edible interior. This is not true tool use, since the gulls do not actually hold or manipulate the breaking surface (Beck 1980), but if there is evidence of cognitive complexity comparable to that of chimpanzee tool use, then shell dropping can confidently be designated as cognitively complex. Furthermore, herring gulls are omnivorous predator/scavengers occupying a very broad ecological niche, and so they should be good subjects in a search for intelligence. I will therefore describe some studies I have done on gull shell-dropping behavior, and compare the cognitive underpinnings of such behavior with the cognitive bases of chimpanzee tool use, which is generally accepted as being cognitively complex (Beck 1982).

Cognitively Complex Tool Use in Gulls and Chimpanzees

I conducted a noninterventive study of shell dropping by herring gulls at Stage Harbor, near Chatham, Cape Cod, Massachusetts. The study comprised 270 hours of observation over 44 autumn days. Over 800 cases of predatory shell dropping were recorded and, since shells are usually dropped more than once, more than 3,000 actual drops were observed. Several species of shellfish were taken. The shells of the larger gastropod species, such as whelks or moon snails, could be occupied by their original inhabitants or by hermit crabs that quickly appropriate vacated gastropod shells (figure 3).

There were two study sites: a seawall, 1 to 3 meters wide and 100 meters long, constructed to control erosion at the mouth of a navigational channel, and two paved parking lots, each 50 by 75 meters. The lots were at the base of a peninsula, 2 kilometers from the wall, which was at the tip of the peninsula. Both the wall and the lots were only about 20 meters from the mid-tide lines of Nantucket Sound.

Both gulls and chimpanzees appear to require internal representations in their respective forms of tool use. Chimpanzees carry their tools up to 90 meters from the place where they select them directly to the place where they will use them. In most cases the point of eventual use is not visible from the point where the tools

Figure 3 A gull drops a gastropod shell to smash it on the rocks below and expose the hermit crab within, which the gull will quickly consume. Such learned behavior may be comparable in its cognitive complexity to chimpanzee termite hunting. (Illustration by Richard Swartz.)

are acquired. It is inferred that the chimpanzees have a mental image of the area that includes both points, and that they can make adaptive preparations with regard to that image. The gulls capture their prey at low tide, from 30 to 200 meters from the wall or the lots. Since the dropping sites are on a peninsula, the point of capture can be at virtually any compass point in relation to the eventual dropping site. Further, since the peninsula is steeply bermed, the point of capture is some five meters below the elevation of either dropping site. Thus the dropping site is not visible and varies in its directional orientation from the point of capture. Nonetheless, the gulls use a distinctive low and fast flight pattern to fly directly to the drop site. They could not see the drop site until they are in its immediate vicinity, when they gain altitude in preparation for the first drop. The gulls, like the chimpanzees, appear to have a cognitive map of the area that includes images of the drop sites, and they can behave adaptively using those images as references.

Jane Goodall (van Lawick-Goodall 1973), William McGrew (McBeath and McGrew 1982), Toshisada Nishida (Nishida and Hiraiwa 1982), Geza Teleki (Teleki 1974), and other observers of chimpanzee tool use were all struck with the selectivity with which chimpanzees choose and make their tools. Termite fishing, for example, requires tools that are sufficiently long, thin, and flexible to be snaked through the nest tunnels. However, they must be thick and sturdy enough to resist breakage. When the end of the tool becomes frayed or bent during use, the chimp may bite off the worn end or may reverse the tool and use the unworn end. Their recognition of the essential qualities of a good tool implies premeditation, a cognitive model, and the ability to strategically adjust behavior to suit prevailing environmental conditions. But shell-dropping gulls are also selective in their choice of "tools." The gull study area around the seawall included the wall itself, hard-packed sand bars and foreshore, and soft vegetated sand dunes. The soft dunes were the most extensive feature, making up over 87 percent of the study area, yet the gulls directed less than 4 percent of their drops toward the soft sand, which is clearly the least appropriate substrate for dropping. In contrast, they directed over 90 percent of their drops toward the wall, which made up only about 1 percent of the study area but which is clearly the most appropriate dropping surface.

Furthermore, the gulls strategically adjusted their altitude for dropping in the parking lots as compared to dropping at the wall. The wall is a narrow structure, demanding considerable accuracy, especially in the commonly stiff winds. Overall accuracy at the wall was 68.5 percent. On the other hand, it is virtually impossible to miss the expansive hard surface of the parking lots, where accuracy was 99.4 percent. The gulls drop from significantly greater heights in the lots than at the wall, presumably because of the different challenge to accuracy. The average dropping height at the lots was about 6 meters whereas the average at the wall was only about 4 meters. The biological significance of this difference is in time and energy consumption. If a hit is assured, it is more efficient to drop from greater heights because there is a higher probability that the shell will break quickly and completely, thereby allowing speedy extraction and consumption of the prey. Fewer drops and shorter time were needed to exploit virtually every type of prey at the lots. That is, where accuracy was assured, the gulls increased their dropping altitude and thereby conserved time and energy. This too implies premeditation and strategy.

Several additional aspects of shell dropping also suggest strategy: (1) When dropping, the gulls usually orient directly into the wind (93 percent of their drops), an orientation that affords maximum aerodynamic stability. In the few wall cases where their heading deviated from the oncoming wind, their accuracy dropped from 68.5 percent to 41.4 percent. (2) Prey that were large and difficult to extract commonly became covered with sand from repeated dropping and manipulation. The gulls often carried sand-covered items to the water and rinsed them, a strategy reminiscent of the celebrated sweet potato washing by the Japanese macaques of Koshima Islet (Kawai 1965). (3) Recall that some shells are occupied by hermit crabs. Hermit crabs lack the hard exoskeleton characteristic of most crabs, and thus must seek the protection of discarded gastropod shells. Their one weapon is a large claw that can deliver a painful pinch. At the wall, where shells would commonly be only incompletely cracked because of the lower drop heights, the gull had to face the claw as it manipulated and extracted the crab. A common first step was to sever the claw and then proceed to extract and eat the now-disarmed crab. In the lots, hermit crab-inhabited shells were usually shattered on the first drop (due to the greater drop height) and the totally exposed and helpless crab would be swallowed claw and all in one gulp. At the wall, the severed claw would later be eaten, in most cases, proving that severing the claw was not an attempt to discard an inedible part.

These examples offer convincing evidence of such cognitively complex, or intelligent, processes as premeditation, formation of strategies, and the flexible use of cognitive maps to respond adaptively with regard to features that are removed from the gull's present surroundings in both space and time. The cognitive bases of chimpanzee tool use in fishing for termites are strictly comparable to the cognitive bases of shell dropping by herring gulls, and both must be viewed as cognitively complex. Any reservations about the imprecise definition and partial redundancy of the names assigned to these cognitive processes would apply equally to the chimpanzees and herring gulls and would not obviate the comparison.

Without elaborating, there is ample evidence that play, practice, and observation are involved in the learning of these behaviors by both chimpanzees and herring gulls (Beck 1982). That the behaviors are learned and not innate makes credible a search for underlying intelligence; that they are learned in the same fashion strengthens the chimpanzee/herring gull analogy.

I trust that we can now disavow any special relationship between tools and intelligence. Indeed, some tool behavior, like that of the chimpanzees, is cognitively complex, but some, like that of the assassin bug and baboon, are either innate or based on simple learning processes. Likewise, behavior not involving tools can be cognitively complex or simple. Tool use may be among the agents selecting for intelligence, but it is neither necessary nor sufficient for intelligence to evolve.

Mental Images, Tool Use, and Intelligence

What then does select for intelligence? Note that the chimpanzees in termite fishing and the gulls in shell dropping are extracting foods that are embedded within an opaque, tough covering: the nest and the shell. The same can be said for Egyptian vultures breaking ostrich eggs, sea otters breaking clams, and many other tool behaviors (Beck 1980). Suzanne Parker and Kathleen Gibson (1977) have correctly identified "extractive foraging of embedded foods" as creating selection for tool use, and in so doing provided a clue to a more general environmental condition that can create selection for intelligence. Both chimpanzees and gulls behave adaptively with regard to features that are removed in space and time: the chimps select tools well before reaching the termite nest at which they will use them, and the gulls orient to a distant dropping site before dropping. Neither the nest nor the drop site can be seen or otherwise sensed at the start of the sequence, just as the termites or edible shell contents cannot be seen or otherwise sensed at the start of termite fishing or shell dropping. It is reasonable to infer that such goals and incentives, be they food items or places, are internally represented in the brain of the chimpanzee or gull. The internal representation need not be visual, and it need not be a faithful (iconic) image of the physical feature. Whatever their nature, the use of internal representations or images of environmental features appears to be a common element of many behaviors that we deem intelligent.

Earl Hunt, writing on the nature of human intelligence, characterizes thinking as "the manipulation of an internal representation of an external environment" (Hunt 1983, p. 142). These representations are stored as knowledge in memory, and their use is not limited simply to recall. They can be mentally manipulated and used to make predictions or to form relationships that can be used to guide behavior intelligently. Note that in Hunt's terms internal representations are not limited to places or to sensory stimulus displays; they can be of a relationship between two individuals (mother-son), a principle (rain follows thunder), or a prediction (if I attack the alpha male he will retaliate). I have phrased these representations of concepts in linguistic terms, but there is ample evidence that animals that lack a language of the human type can form such concepts, that is, the representations need not be linguistic.

I have not done justice to the current vitality of research on the formation and manipulation of internal representations. It is an old idea, but workers in many disciplines are converging empirically on internal representations (Roitblat 1982), and it promises to be a synthesizing concept for the study of intelligence in humans and nonhumans. I am proposing that tool use will be intelligent when learned patterns of tool use are predicated on the formation and manipulation of internal representations. Many other behaviors, not involving tools, are also predicated on internal representations and will also be intelligent.

Tool Use and Intelligence in Hominid Evolution

Our own species, *Homo sapiens,* has one of the largest brains of any species and, perhaps not surprisingly, on our own tests of intelligence scores as the most intelligent species. In addition to our intelligence, humans make the greatest number and variety of tools and are the most dependent on tools for survival. The existence of great intelligence and great tool proficiency in humans has been taken by many as evidence for a causal relationship between the two. Furthermore, the human brain first began to expand in size and complexity about two million years ago, at the same time that manufactured stone tools first appear in the archeological record. However, the proliferation of stone tool culture and the growth of the human brain and intelligence were probably both propelled by a third factor: evolution of an advanced capacity to form and manipulate internal representations. We infer this capacity from reconstructions of the lifeways of early hominids.

About two million years ago, hominids became committed to a life of gathering, scavenging, and hunting. Hunting, especially coordinated hunting of game, involves planning and communicating about events that would occur days later and miles away. Gathering requires a vast knowledge of the identity, location, and seasonal availability of many different plant and animal products. These early humans (*Homo habilis* and *Homo erectus*) occupied a broad subsistence niche, that of a generalist omnivore, which we have already seen to create selection for intelligence. The change in subsistence was accompanied by demographic and social changes that also favored intelligence. These people lived longer and had longer periods of childhood, presumably to learn the information needed for such complicated subsistence. They lived in stable social groups and occupied base camps for long periods. Economic cooperation was essential. It is generally assumed that females, children, and seniors did the gathering, operating within a day's round trip of the home base. Males, unencumbered by infants and muscularly more robust, hunted and may have been absent from the rest of the group for several days. Hunters and gatherers, to survive fluctuations in food abundance, would have had to pool and share their harvest and prey. Such economic cooperation involved food-sharing between bonded males and females, cooperative hunting and mutual assistance, and solicitous provisioning of the ill, injured, and aged. These interactions created selection for the ability to memorize and utilize such abstract social information as kinship, reliability and skill of partners, and indebtedness. Tool making, in addition to requiring extensive knowledge of materials and techniques, also required the ability to conceptualize a functional end product when viewing a lump of stone.

This reconstruction suggests that hominids embarked on a lifeway about two million years ago that necessitated the acquisition, storage, manipulation, and recall of unprecedented amounts of knowledge. This knowledge was stored as internal representations of an external social and physical environment. Selection intensified for the ability to store and use internal representations, an ability that may be synonymous with intelligence. Eventually hominids acquired language, a symbolic system that greatly facilitates the storage, manipulation, and communication of internal representations.

In summary, some, but by no means all, of the tool behavior of nonhuman animals provides evidence for cognitive complexity. However, many behaviors that do not involve tools appear to be equally intelligent. Although tool proficiency and brain size evolved in concert in the human lineage, tools were only one of many factors that created selection for the ability to store and use knowledge in the form of internal representations. We must discard the notion of any simple or unique causal relationship between tools and intelligence, either in living species or in extinct human ancestors.

Select Bibliography

Beck, B. B. 1980. *Animal tool behavior.* New York: Garland STPM Press.

———. 1981. Chimpocentrism: Bias in cognitive ethology. *Journal of Human Evolution* 11:3–17.

Hall, K. R. L. 1963. Tool-using performances as indicators of behavioral adaptability. *Current Anthropology* 4:479–94.

Hunt, E. 1983. On the nature of intelligence. *Science* 219:141–46.

Kawai, M. 1965. Newly acquired pre-cultural behavior of the natural troop of Japanese monkeys on Koshima Islet. *Primates* 6:1–30.

McBeath, N. M., and W. C. McGrew. 1982. Tools used by wild chimpanzees to obtain termites at Mt. Assirik, Senegal: The influence of habitat. *Journal of Human Evolution* 11:65–72.

McMahan, E. 1983. Bugs angle for termites. *Natural History* 92:40–47.

Nishida, T., and M. Hiraiwa. 1982. Natural history of a tool-using behavior by wild chimpanzees in feeding upon wood-boring ants. *Journal of Human Evolution* 11:73–99.

Parker, S., and K. Gibson. 1977. Object manipulation, tool use and sensorimotor intelligence as feeding adaptations in cebus monkeys and great apes. *Journal of Human Evolution* 6:623–41.

Roitblat, H. L. 1982. The meaning of representation in animal memory. *Behavioral and Brain Sciences* 5:353–406.

Teleki, G. 1974. Chimpanzee subsistence technology: Materials and skills. *Journal of Human Evolution* 3:575–94.

Van Lawick-Goodall, J. 1973. Cultural elements in a chimpanzee community. In *Precultural primate behavior*, ed. E. Menzel. Basel: Karger, pp. 144–84.

Tradition and Social Learning in Animals

Bennett G. Galef, Jr.

Tradition and social learning are intimately related because the capacity for social learning is a prerequisite for the establishment and maintenance of true traditions in a population. Although tradition and social learning are clearly linked to one another, the methods appropriate for the study of each are quite different.

A tradition is a learned pattern of behavior that is common in a particular social group, but absent in other social groups of the same species. The existence of a tradition can be identified by careful, unobtrusive observation of organisms in undisturbed environments. The term "social learning," on the other hand, refers to a class of behavioral mechanisms that may result in the production of a tradition in a population. Such mechanisms can be analyzed only by experimentation under controlled conditions.

The reason why this distinction between tradition and social learning is important is that observation of a tradition in a population of animals has frequently been used to infer the existence of a complex social learning process underlying that tradition. I intend to show that such inferences are not justified, and that traditions may be established and maintained in populations by quite simple types of social learning. The existence of traditions in a species is not necessarily evidence of an ability of members of that species to learn by imitation, observation, or in any other sophisticated way. For example, it has been well established by observation that many species of British birds have acquired the habit of opening milk bottles and eating the cream from the milk's surface (Fisher and Hinde 1949).

Description of the spread of this habit among birds in several areas strongly suggests that social interaction is important in propagation of the behavior. But is it the case that one bird learns to open milk bottles by watching another bird do so? It is perhaps at least as likely that after one bird has opened a milk bottle, others come to feed from the opened container and are themselves subsequently more likely to attempt to feed at closed bottles. The existence of a tradition of feeding from milk bottles is well established, but it is not clear what inferences concerning the learning capacities of birds should be drawn from the existence of that tradition.

A Brief History of the Study of Animal Traditions and Social Learning

The possibility that traditions may rest on rather humble foundations was first suggested more than 70 years ago by Edward Thorndike (1911), the founder of North American animal experimental psychology. However, Thorndike's approach to the analysis of social learning has not as yet, at least not in many popular texts, replaced the view prevalent in the late nineteenth century that traditions are indications of considerable mental sophistication in their bearers. It is worthwhile to recount some of the relevant history of the issues involved in order to clarify some of the confusion that still exists today.

In the decades before comparative psychology or the study of animal behavior from a biological perspective became experimental disciplines, it was commonly believed by naturalists, pet owners, animal trainers, and others familiar with animals that mammals generally could learn to perform complex acts by observing and then imitating others performing those acts. There was no compelling evidence of learning by observation or imitation in animals; belief in such processes was just one of a number of widely held but unexamined notions about animal functioning common during the Victorian era.

The question of whether animals would, in fact, learn by observation became theoretically important at the close of the nineteenth century as the result of a fundamental disagreement between the co-originators of evolutionary theory, Alfred Russel Wallace and Charles Darwin. Though Wallace and Darwin agreed about many things, they differed fundamentally over the origins of the human mind. Darwin was convinced that the human mind, like other features of living organisms, was the product of purely natural processes. Wallace argued that natural processes alone were insufficient to produce the intellectual sophistication of mankind. Thus, in the 1880s, when the question of the applicability of evolutionary models to understanding of human functioning was actively debated (as it is today in a different context), the question of whether animals had faculties of mind similar to those found in man was of importance. If one could identify simplified precursors of human intellectual capacities in animals, then continuity from animal to man in intellectual life would be established and there would be no need to invoke extra-natural causes in discussing the origins of human mental functioning. If, to the contrary, man had intellectual capacities for which no simpler precursors could be found in nonhuman animals, then an evolutionary explanation of the human mind would be more difficult to maintain.

The major contributor of evidence to the Wallace-Darwin controversy was Darwin's protégé and disciple, George Romanes, a Fellow of the prestigious Royal Society and a leading figure in the biological establishment of his day. Romanes's approach to the problem of determining whether there was continuity in the intellectual capacities of animals and man was to postulate a hierarchy of mental faculties (Romanes 1882, 1884). This linear scale of qualities of mind extended from the protozoa, which it was said exhibited only excitability, conductility, and the capacity to discriminate among stimuli, to modern Western man, the possessor of the faculties of reason, conscience, and abstraction in their highest states of development. Using such a scale one could, at least in principle, rank-order species by their capacity to exhibit the mental faculties most clearly shown by humans.

Imitation was an important mental faculty for Romanes's (1884) analysis because, he argued, the capacity to imitate (as well as several other higher mental faculties) was not unique to the most highly evolved form, adult Western man, but could be found elsewhere in the phylogenetic scale. In fact, Romanes claimed, imitation reaches its highest level of perfection in slightly inferior forms: monkeys, children, savages, and idiots. So, if one could find evidence of learning by observation or imitation in nonprimate animals, one would be providing strong evidence of a continuity of mental faculties and hence providing support for the Darwinian notion of the evolution of human mind by natural processes.

George Romanes was a strong proponent of the view that learning by observation is central to behavior acquisition in animals. In 1882, Romanes published an influential monograph, *Animal Intelligence*, in which he provided more than five hundred pages of anecdotal description and interpretation of instances in which animals exhibited rather remarkable intellectual powers in the solution of problems they encountered in their natural environments. However, many of the examples of animal learning in nature reported by Romanes and his correspondents were fanciful at best. For example, mice in Iceland were said to have been observed storing supplies of berries in dried mushrooms, loading these rations onto dried cow-droppings, and then launching their improvised, provisioned vessels and guiding them across rivers using their tails as rudders. It was assumed that the mice had originally acquired these abilities by observing and imitating humans, and that the capacities to store provisions and construct and steer rafts had become traditional in some mouse populations. However, not all of the evidence of intelligence in animals described by Romanes was quite so unlikely, and some was central to future developments in the study of the learning capacity of animals.

Possibly the most historically important of Romanes's many cases of supposed "imitation learning" concerned a cat that belonged to his coachman. This animal had learned, without formal tuition of any kind, to open a latched door in Romanes's yard by jumping up and grabbing the latchguard with one forepaw, depressing the thumbpiece with the other forepaw, and simultaneously pushing at the doorpost with her hind legs. Romanes argued that the cat, in the absence of any other source of information, must have observed that humans opened the door by grasping the handle and moving the latch. Then, said Romanes, the cat must have reasoned, and I quote, "If a hand can do it, why not a paw?" Finally, strongly motivated by this insight, the cat attempted to and succeeded in opening the door.

The problem with Romanes's interpretation, of course, is that simple observation of an animal behaving in an uncontrolled environment provides little useful information about the processes responsible for the development of the animal's behavior. It is impossible to tell from simply watching an animal perform an act in an uncontrolled setting what the necessary antecedent conditions of that performance are.

Experimental animal psychology in North America may well

have arisen out of Edward Thorndike's irritation with the excesses in Romanes's *Animal Intelligence:* its anecdotal method, its speculative conclusions, and most particularly its insistence on the importance of observational learning and tradition in the development of the behavior of animals. In the late 1890s Thorndike brought the door-opening behavior of cats, described by Romanes, into the laboratory and studied the acquisition of solutions to a variety of mechanical problems in controlled and replicable situations. In one of Thorndike's experiments, food-deprived cats were individually placed in a wooden cage and observed as they learned to depress a treadle located in the center of the floor of the apparatus in order to escape confinement and gain access to a food bowl placed outside the cage (figure 1). As is well known, on the basis of the results of his

Figure 1 One of Edward Thorndike's puzzle boxes, designed for the study of trial-and-error learning in cats. When the cat steps on the treadle in the center of the floor, the door opens and the cat can escape confinement and gain access to food. (Illustration by Richard Swartz, based on a drawing supplied by B. G. Galef, Jr.)

studies in a variety of such puzzle boxes, Thorndike (1911) proposed that animals learn to solve *all* such problems, including presumably the opening of garden gates, as the result of their *individual interactions* with the environment, by a gradual process of trial-and-error learning. Less generally appreciated is Thorndike's explicit rejection, on both theoretical and empirical grounds, of the possibility of learning by observation. Thorndike had found that animals in general, and cats in particular, did not learn to get out of puzzle boxes either by observing other cats do so or by observing humans demonstrate solutions. In fact, Thorndike found that observation of a trained demonstrator by a naive individual sometimes interfered with the gradual process of trial-and-error learning by which naive individuals acquire solutions to a variety of problems.

Of course, it cannot be inferred from the lack of evidence for true imitation learning that other sorts of social learning might not be important to animals in their natural environments. (Thorndike himself was careful to point out that what he called "semi-imitative" phenomena, the "indirect results of instinctive acts" of various kinds, could accelerate learning.) In fact, there are certain behavioral phenomena in nature that appear to *require* explanation in terms of social learning of some kind—behavior patterns that are usually referred to as "traditional" (Galef 1976). If you compare the behavior of members of a single species living in nature in different social groups, as any number of field biologists have done, you will often find that many of the members of one social group will exhibit a pattern of behavior totally absent in other groups. Such intergroup differences in behavior are most commonly observed in patterns of food selection or in the motor patterns involved in food acquisition.

Field biologists, observing such intergroup variation in behavior, have long assumed that such animal traditions are transmitted from individual to individual within a group by observational learning or imitation. However, data from many psychology laboratories over many years suggest that observational learning and imitation are not very important processes in behavior acquisition, at least not in nonprimates. So there remain important questions as to the processes supporting the development and maintenance of the "traditional" patterns of behavior to be observed in many nonprimate vertebrate social groups.

Experimental Analysis of the Learning of a Tradition in Wild Rats

For the past decade my students and I have been studying the role of social process in the development of traditional patterns of behavior in wild Norway rats. In the next few pages I will describe one of several instances in which we have attempted to determine the causes of idiosyncratic feeding patterns exhibited by our animals. We began with field observation of a traditional pattern of behavior, brought the phenomenon into the laboratory, and then attempted to analyze the social learning mechanisms supporting that behavior (Galef 1982; Galef and Clark 1971).

Some years ago an applied ecologist, Fritz Steiniger, was working for the German government as a rodent control officer. He noticed a rather peculiar thing. Steiniger found that if he employed some poison bait in an area for an extended period of time, he would have considerable initial success, with the rats eating lots of posion and dying in large numbers. Later, however, acceptance of the bait was very poor. Steiniger noted in particular that young rats born to adults that had survived poisoning rejected the poisoned bait without ever even sampling it themselves. These young fed exclusively on safe diets available in their colony territory and totally avoided contact with the poison bait their elders had previously learned to avoid.

This is a robust phenomenon and relatively easy to capture in the laboratory. In our basic experiment, we established colonies consisting of two male and four female adult wild rats in 3-by-6-foot enclosures each containing four wooden nest-boxes. Water was continuously available and food was presented to the colony for 3 hours a day in two food bowls located about 2½ feet apart. Each bowl contained one of two nutritionally adequate diets, each discriminable from the other in color, texture, taste, and smell. For simplicity, I will refer to these two diets as diets A and B in all that follows.

The adult members of our colonies were trained to eat one of the diets presented each day and to avoid the other because it was laced with lithium chloride, an illness-inducing agent.

Under these conditions our wild rats rapidly learned to avoid eating the contaminated diet and, most important, continued to avoid the previously contaminated diet for some additional weeks when they were offered uncontaminated samples of it. So we have colonies of adult wild rats eating either diet A or diet B, and avoiding the alternative because of its previous association with illness.

The experiments proper began when litters of pups that were born to colony members left their nest-site to feed on solid food for the first time. We observed the adults and pups throughout daily 3-hour feeding periods on closed circuit television and recorded the number of times the pups ate from each of the two food bowls, now containing uncontaminated samples of diet A and diet B. We found that pups born to a colony trained to avoid eating diet B ate only diet A, the diet that their parents had been trained to eat. Pups born to a colony trained to avoid diet A ate only diet B and never even made contact with diet A (figure 2). Observations of more than 240 wild rat pups during their first two weeks of feeding on solid food have revealed only a single individual that ate any of the diet that the adults of its colony had learned to avoid.

After a litter of pups had been feeding on solid food for two weeks, we transferred them to a new enclosure, similar in size but different in layout from their original home. Here, without the adults of their colony, the pups were again offered a choice between uncontaminated samples of diets A and B. The amount of each diet eaten

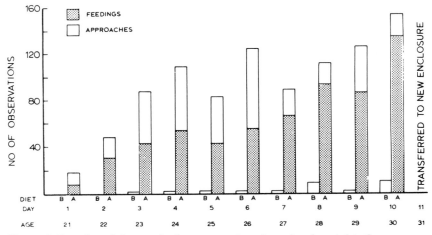

Figure 2 Results of the study demonstrating the role of social influence on diet selection in weanling wild rats. Although food B is normally preferred by rats, the adults learned to avoid it when it was experimentally altered to make them ill. The graph shows that weanling rats, who never experienced illness associated with food B, also showed the adult pattern of rarely approaching and never eating food B.

by the pups in this situation was determined by weighing food bowls before and after each feeding session. We found that the pups continued for 8 to 10 days to prefer the diet that the adults of their colony had eaten, even though the pups were now living and feeding without contact with those adults.

Taken together, these observations demonstrate, as Steiniger suggested, that adult rats can, in some fashion, lead their offspring to feed solely on a safe diet in an environment containing food known by the adults to have been poisoned. The data also show that the food preferences learned in the presence of adults continue to affect the diet preference of pups for some time after their removal from direct adult influence. So there can exist traditions in the food preferences of colonies of wild rats. The important question is how are such traditions established and maintained in a wild rat population?

Over the last few years, my students and I have found a number of ways in which adult wild rats can induce their young to wean to a given food. For example, we have found that the physical presence of adults at a feeding site attracts pups to that feeding site and markedly increases the probability of young rats weaning to the food located there. If one establishes a colony of adult wild rats in a large enclosure (12 by 8 feet) makes diet A continuously available in two food bowls located 10 feet from the nesting area, and continuously (24 hours a day) monitors behavior at the food bowls, one can determine the conditions under which each individual pup in a litter eats it first meal of solid food (figure 3). We have observed nine individually marked pups from three litters take the first meal of solid food and all nine ate their first meal under exactly the same circumstances. Each ate its first meal while an adult was eating and each ate at the same food bowl as the feeding adult, not at the other food bowl 1.5 feet away. Given the observed temporal and spatial distributions of adult meals, the probability of those conditions occurring nine times in succession by chance was very small indeed, less than four in a thousand. So the presence of an adult at a feeding site serves to attract pups to that site and to cause pups to initiate feeding there.

In addition to being able to influence a pup's choice of feeding site, and thus indirectly its food preference, the mother of a litter of pups can also directly influence her own pups' dietary preference. We have conducted an experiment much like the first one described above, but with one important difference. Colonies of adult rats were again housed in 3-by-6-foot enclosures, but adults were removed to a

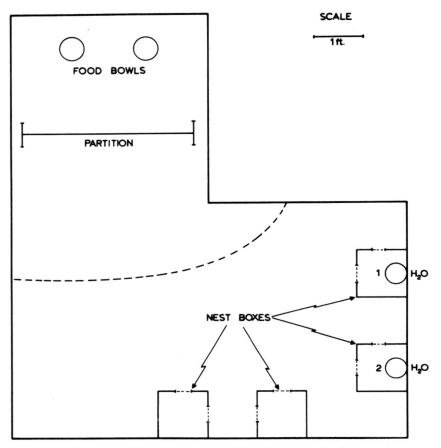

Figure 3 Floor plan of the enclosure used for observing the first meals eaten by wild rat pups. The area above the dashed line is continuously monitored on closed circuit television. Adults established nests for their young at 1 and 2.

separate cage for 3 hours each day, where they were fed either diet A or diet B, depending on the experimental condition to which their colony was assigned. While the adults were out of the colony enclosure, the pups were presented with two standard food bowls, one containing diet A and the other diet B. We found that the diet eaten by the adults profoundly affected the food choice of the pups even though the adults and young had no opportunity to interact directly in a feeding situation. Once again pups from colonies of adults trained to eat diet B ate diet B, while those from colonies of adults trained to eat diet A preferred that diet.

Our research indicates that the milk of a lactating female rat contains cues directly reflecting the flavor of her diet. We believe that at weaning, as the result of prior exposure to these flavor cues, rat pups exhibit a preference for a diet of the same flavor as the diet that their mother had been eating during lactation (Galef and Sherry 1973).

Fritz Steiniger was right. The learned feeding preferences of adult wild rats can be socially transmitted to their young, reducing the probability that the young will ingest toxic food.

Edward Thorndike was also right. The indirect results of what might be conceived of as instinctive acts, in this case the tendency of rat pups to approach adults and to suckle from their mother, can result in introduction of the young to their parents' diet and consequent *apparent* imitation of learned adult food preferences by the young. So here we have a case in which an animal tradition rests not on imitation or observational learning, but instead on some rather simple sorts of exposure learning.

Social influences on foraging behavior and diet selection are important to rats and are supported by a range of simple but elegant social learning processes. One of these that we are analyzing now is quite powerful. Anyone seeing rats in the wild exhibiting the behavior I am about to describe might be convinced that they are very clever indeed. The explanation, however, is really quite simple.

The laboratory procedure we used was designed to mimic a situation in which a foraging rat ingests a food item at some distance from its burrow, returns to the burrow, and then interacts with a familiar burrow-mate that subsequently selects a food item to eat. We were interested to know whether, as the result of such interaction at a distance from a feeding site, a burrow resident could acquire information concerning the food a returning forager had eaten (Galef and Wigmore 1983).

In brief, our procedure involved feeding one rat (a demonstrator) one of two novel-tasting diets, either cocoa-flavored diet or cinnamon-flavored diet, and then allowing the demonstrator to interact with a second rat (an observer) for 15 minutes. Immediately following this brief period of interaction between demonstrator and observer, the demonstrator was removed from the experiment. Then, for 60 hours, the observer was offered a choice between two food cups, one containing cinnamon-flavored diet and the other containing cocoa-flavored diet. We weighed the food cups every 12 hours and found, much to our surprise, that observers whose demonstrators had eaten cocoa-flavored diet ate much more cocoa-flavored than cinnamon-flavored diet, while observers whose demonstrator had eaten cinnamon-flavored diet ate much more cinnamon- than cocoa-flavored diet. The effects of the demonstrators' diet on the observers' diet preference were still very strong 60 hours following interaction of observer and demonstrator.

Rats can communicate to one another information concerning diets ingested at a time and place distant from the locus of communication. Further, information passing from demonstrator to observer concerning the food the demonstrator has eaten has profound effects on the subsequent food preferences of observers. How do the rats influence one another's food preferences in this way? Quite simply, when an observer rat is exposed to the smell of a food item on the breath of a demonstrator rat the observer subsequently exhibits a strong preference for the food eaten by the demonstrator.

My students and I have developed several converging lines of evidence each of which is consistent with the hypothesis that olfactory cues passing from demonstrator to observer cause observers to prefer their demonstrator's diet. I will describe two of these lines of evidence very briefly.

If, after the demonstrator has eaten either cocoa- or cinnamon-flavored diet, and before it interacts with an observer, the demonstrator is anesthetized, taped to a stand, and placed for 15 minutes with its nose 2 inches from a screen that separates the sleeping demonstrator from its observer, the message still gets through; the observer interacting with a sleeping demonstrator 2 inches away still exhibits during testing a strong preference for that demonstrator's diet (figure 4). Which tells us two things: first, the effective message is emitted in a passive way by the demonstrator and is not elicited by the observer; and second, no physical contact between demonstrator and observer is required for information transfer to occur. The important cue can be transmitted over some distance; therefore, it is not a taste cue, which strongly suggests that olfactory cues (smells) are carrying the message.

In another experiment, we found that if we render an observer anosmic (unable to smell) by rinsing its nasal cavity with zinc sulfate solution prior to the time that an observer interacts with its demonstrator, the observer subsequently fails to exhibit a preference for its demonstrator's diet during testing. Control observer rats, whose nasal passages have been rinsed with a neutral saline solution

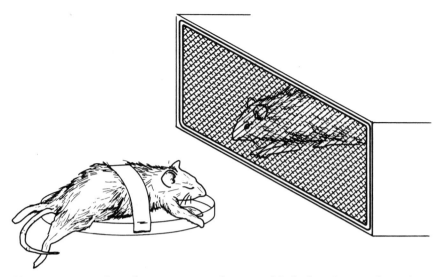

Figure 4 An awake "observer" rat with no established preference for either of the strong-smelling experimental foods is exposed to the sight and odor of an anesthetized "demonstrator" rat who has eaten one of the foods. The observer will later show a strong preference for the food eaten by the demonstrator, indicating that food preferences can be transmitted passively and without physical contact.

prior to their interaction with demonstrators, do show a strong tendency to eat the same diet that their demonstrator has eaten. Thus, olfactory sensitivity in observers is necessary for information transfer to occur.

Once again we have *apparent* imitation of one rat by another resting on a very simple social learning process. Smelling a food on the breath of a conspecific induces a preference for that food and apparent imitation of demonstrators by observers.

Conclusions

Several well-known examples of animal traditions described in the literature have not yet been analyzed in detail. I am sure that many people have heard of the monkeys of Koshima Islet in Japan that clean sand from the skin of sweet potatoes by dipping the potatoes in water before eating them, or have read about the troop of monkeys whose members have learned to sort wheat from sand by throwing handfuls of the mixture onto water. Then there are Jane Goodall's chimpanzees in the Gombe reserve that fish for termites using twigs as tools, and several species of British birds that, as I mentioned earlier, have learned to open milk bottles and eat cream from the surface of the milk. While there is no doubt about the reality of such behavioral phenomena, it is premature to assume that such patterns of behavior pass from one individual to another as the result of imitation or observational learning. There is no reason to believe that such traditional behaviors are evidence of any particularly great intellectual prowess in those animals that exhibit them. The fact of the existence of a traditional behavior pattern does not tell us anything about how the traditional behavior was acquired or transmitted.

It is important to keep in mind that simple acquisition processes can be responsible for rather complex behavioral outcomes. Until the processes of acquisition of such traditions can be examined in detail under controlled conditions, they remain thought-provoking observations, not evidence of the reality of special mental abilities in those creatures that exhibit traditions. A healthy skepticism and a commitment to empiricism are necessary for the development of understanding of the social learning processes resulting in traditions in animals.

Select Bibliography

Fisher, J., and R. A. Hinde. 1949. The opening of milk bottles by birds. *British Birds* 42:347–57.

Galef, B. G., Jr. 1976. Social transmission of acquired behavior: A discussion of tradition and social learning in vertebrates. In *Advances in the study of behavior*, ed. J. S. Rosenblatt, R. A. Hinde, E. Shaw, and C. Beer. Vol. 6. New York: Academic Press.

———. 1982. Studies of social learning in Norway rats: A brief review. *Developmental Psychobiology* 15:279–95.

Galef, B. G., Jr., and M. M. Clark. 1971. Social factors in the poison avoidance and feeding behavior of wild and domesticated rat pups. *Journal of Comparative and Physiological Psychology* 75:341–57.

Galef, B. G., Jr., and D. F. Sherry. 1973. Mother's milk: A medium for the transmission of cues reflecting the flavor of mother's diet. *Journal of Comparative and Physiological Psychology* 83:374–78.

Galef, B. G., Jr., and S. W. Wigmore. 1983. Transfer of information concerning distant foods: A laboratory test of the "information-centre" hypothesis. *Animal Behaviour* 31 : 748–58.

Romanes, G. J. 1882. *Animal intelligence.* London: Kegan Paul, Trench.

———. 1884. *Mental evolution in animals.* New York: Appleton-Century-Crofts.

Thorndike, E. L. 1911. *Animal intelligence.* New York: Macmillan.

Do Animals Think?

Carolyn A. Ristau

Charles Darwin speculated about animal mental experience in his book of 1872, *The Expressions of the Emotions in Man and the Animals*, but it was George Romanes, a close friend and protégé of Darwin's, who first explored the issue of animal mind in depth, in his books *Animal Intelligence* (1882) and *Mental Evolution in Animals* (1884). Although the intellectual interests of Romanes were basically those of this symposium—to understand and analyze the intellectual abilities of different animals and to compare them with those of men—his methods were faulty. He was too accepting of anecdotes and tended to anthropomorphize excessively, though both anecdotal evidence and an anthropomorphic approach can be useful to scientific progress.

C. Lloyd Morgan (1894), like Darwin, also believed in a gradual evolution of mental abilities. He emphasized the need for simplifying explanations, and is well known for his oft-quoted phrase, "In no case may we interpret an action as the outcome of the exercise of higher psychical faculty, if it can be interpreted as the exercise of one which stands lower on the psychological scale." Morgan's "Law of Parsimony," as the rule is often called, was a *modus operandi*, a heuristic device, an attempt to allow research to proceed efficiently and effectively. Morgan, like scientists today, recognized that Mother Nature is not simple. His followers, however, as is usually true of movements that arise in the wake of pioneer thinkers, tended to simplify and exaggerate.

As scientific interest in the study of behavior continued, there was continued interest in mental experience, especially in human mental experience. E. B. Titchener, Wilhelm Wundt, and others sought to understand the mind by the process of introspection.

Titchener was a reductionist who tried to reconstruct mind from minimal perceptual elements such as bits of color and light that could be isolated from mental experience by introspection.

But soon an opposing swing took hold, one that avoided all mentalistic terms and sought to "explain" or at least to predict human and animal behavior in terms of specifiable stimuli and responses. That movement was and is behaviorism, and names like B. F. Skinner became widely known. It too had its failings—most easily seen in Skinner's attempts to understand thinking and human language use exclusively in terms of stimuli and responses.

In recent years, scientists have become concerned once again with mental experience, but this time they are equipped with a variety of approaches and are armed with the more sophisticated technology of our day and experience in behavioral experimental design. They have some of the methods of the behaviorists, but ask different questions. In this chapter, I will discuss some aspects of this contemporary approach to animal thinking.

Why might we think that animals think?

If animals do not think, what is it they could be doing instead? Are animals mere stimulus-response machines, in effect responding without perceiving, not having awareness of what it is they are doing? Should we consider an animal merely as a kind of box into which specifiable stimuli are put and out of which predictable responses emerge?

Yet what do we mean when we say an animal is aware of something or knows what it is doing? We mean that an animal does not merely react to a stimulus, perhaps some food before it, as a machine might. Being aware of the food means it visualizes the food, it smells the food, it has the experience of smelling.

Before we begin struggling to define thinking and other kinds of mental experiences, let us consider why we might be led to consider that animals think.

For one, continuity between man and other animals might lead us to think animals think. The history of evolution provides us with many examples of ways in which various animals species, including man, are similar to each other. Man shares enormous similarities with other animals, though, of course, with important differences.

We find similarities in the anatomy of the respiratory, cardiovascular, and other systems, as well as bilateral symmetry, which we share with so many species. Our chemistry exhibits evolutionary continuity. Even behavior is similar; we and many other animal species form social groups, we communicate with each other, we care for our young, we act aggressively toward each other. Why not then mental continuity as well? Indeed, we and most existent animal life have nervous systems composed of neurons that are remarkably similar, and most neurons are connected via synapses that also seem quite similar. If we wish to assume other species are not conscious, what is it that is so distinctly different about our nervous system that precludes consciousness in other species? In brief, we exhibit evolutionary continuity with animals in so many other ways, why not mentally as well?

A second reason we might think animals think is that animals, like ourselves, face difficult problems existing in the real world for which thinking could be very useful. Many of them must escape predators. They must hide or run away, often in devious ways. Alternatively, some animals have to find and hunt prey, sometimes apparently hunting cooperatively. Before encountering obstacles, it would be helpful to think ahead of a good detour around. The ability to think and reason could be quite useful in all these situations.

Thinking might also be *easier* than some other possible means of dealing with many situations. Let us consider specifically the possibility that some species of animals might be able to consider that another had a certain feeling or emotion, such as liking. From the point of view of both the animal having to deal with another and of the scientists trying to interpret the animal's behavior, it might be easier to consider that animal A *likes* animal B, rather than to make a long and complicated series of specific behavioral predictions. The making of specific behavioral predictions is the way most of us have been trained to do science. For example, if we were looking at a group of monkeys or chimpanzees, which is simpler: (1) to keep in mind simultaneously such bits of information as monkey A sat near to monkey B at a distance of 6 inches or less on the following X number of occasions, and monkey A huddled next to monkey B on Y number of occasions, and in fact on Z number of those occasions had its tail next to monkey B, which from correlations we have made is related to the likelihood of monkey A allowing monkey B to carry its infant, and so forth; or (2) to assume that monkey A *likes* monkey B, based of course on these same data which, however, we need not remember in detail, only as they weigh in for or against the idea that monkey A likes monkey B. Having this concept of liking we and the monkeys might be able to make predictions about behaviors not previously exhibited to us by these two monkeys. For instance, if monkey C began attacking monkey B, both we and the monkeys could make the reasonable prediction that monkey A will come to the assistance of monkey B in defending it against C. The alternative, remember, is to hold in mind simultaneously all the other specific behaviors and then to make a prediction about a behavior we have not seen exhibited by this particular hypothetical pair, namely assisting in an aggressive encounter. Briefly, even something so advanced as attributing mental states to another may be simpler than the traditional ways experimental psychologists have felt that scientists, and the animals as well, should be making their predictions.

What are some of the possible realms of mental experience for an animal?

We have so far glossed over problems in defining thinking or other mental experiences. Let us now consider a list that Donald Griffin (1981, 1982) has suggested as possible mental experiences, ranging from those generally considered "O.K." to use with respect to animals, or perhaps even humans, to those considered to be "taboo" (see figure 1). Different scientists might order the list differently, and would draw the line between what it is scientifically valid to speak of and what it is not, at different places along this list.

For the present purposes, we are most interested in the terms close to the taboo side. To begin, we can distinguish between different possible aspects of "awareness" as various philosophers have done. Perhaps at a quite rudimentary level "awareness" could mean "having a sense of" an external object such as a rock, or energy, including gravity and various frequencies of electromagnetic radiation, such as visible light and ultraviolet light. We could extend the notion of "awareness" to awareness of mental states.

"Thinking" is a term sometimes used to refer to almost any mental experience. As others have suggested (Pylyshyn 1978), thinking involves the ability to represent not only objects and events that are occurring here and now but also ones remote in space and time, a property known as displacement. Thinking is more than mere repre-

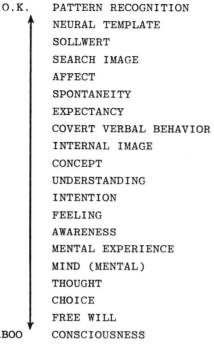

```
O.K.     PATTERN RECOGNITION
  ↑      NEURAL TEMPLATE
         SOLLWERT
         SEARCH IMAGE
         AFFECT
         SPONTANEITY
         EXPECTANCY
         COVERT VERBAL BEHAVIOR
         INTERNAL IMAGE
         CONCEPT
         UNDERSTANDING
         INTENTION
         FEELING
         AWARENESS
         MENTAL EXPERIENCE
         MIND (MENTAL)
         THOUGHT
         CHOICE
  ↓      FREE WILL
TABOO    CONSCIOUSNESS
```

Figure 1 A gradient of possible mental experiences ranging from those generally accepted in regard to animals to those that are "taboo" and usually used only in reference to human beings. Recently, some investigators of animal behavior have become more willing to apply terms at the taboo end of the scale to animals. (From Griffin 1976.)

sentation; as typically used, thinking involves manipulation of representations, some reordering of them, possibly in ways we have not directly experienced. I like the phrase used by Gould and Gould (this volume), that thinking is "cognitive trial and error." Benjamin Beck (this volume) also defines cognitive behavior in similar ways.

There are, as you may imagine, a number of disputes about the nature of thinking. Some philosophers consider that thinking requires words or language. Clearly if this is the case, and if the language they mean is much like human language, then we are not going to find much thinking in the animal world. If, however, we allow that thinking can be done without words, as many other philosophers seem to believe, then at least we can consider the possibility of thinking in nonhumans.

Another problem that arises is whether thinking implies consciousness. Many philosophers have long separated the two, and many contemporary cognitive psychologists are content to study cognitive processes in both humans and animals without making any assertion that such processes need be conscious. Indeed, some human thinking does appear to be unconscious, and it would seem likely that some animal thinking is unconscious, too.

Strictly speaking, then, to show that an organism requires thinking to solve a particular problem does not necessarily imply consciousness as well. But we are hard put to allow only ourselves conscious sensory experiences and thought in problem solving and to deny the same in a nonhuman organism that solves similar problems. If we deal only with necessary and sufficient conditioning of consciousness we are left in the solipsist's position: "I can know only that I am conscious for I can know only my own mental states. I cannot know whether you or anyone else is conscious." But even most philosophers do not leave themselves in the solipsistic state. I hope the reader will wonder about the adequacy of "species solipsism" as well (Griffin 1984).

Supposing, then, if an animal can think, what might we expect it to think about? A homing pigeon in flight might think about what it is doing up in the air—looking for food or going to the homing site? When building a nest, it could be useful for a bird to have some notion of what the finished nest will look like, rather than only following a motor pattern of how to put twigs together. For various species, researchers have poked holes in nests in an effort to see how rigidly programmed the birds might be in their nest building. Generally speaking, the birds were able to repair a hole even though that was part of the nest they had already built, suggesting at least some notion of what the nest should look like. When animals are choosing a territory, they might think about what attributes they are choosing. As Griffin (1984) asks, when birds choose a territory, which in several months should have a plentiful supply of insects to feed their young, but which now contains only insect eggs or larvae, how are the birds making that choice? And of course, as I have already mentioned, for animals with fairly complex social groups, thinking would be very useful in manipulating and predicting social interactions. Nicholas Humphrey (1980) and Alison Jolly (1966) have previously made this suggestion, and Colin Beer (this volume) also discusses the social basis of intelligence. The list of matters that animals could think about is quite endless.

Some Examples from the Laboratory and the Field that Are Suggestive of Thinking in Animals

Studies of the ability to form categories

The ability of animals to make discriminations not only between particular items but between classes of similar items is important for getting along in the world. In the laboratory there are many discrimination learning tasks that animals of various species can do. Usually these tasks require the animal to respond differentially to different tones or to geometric figures of different sizes, shapes, or colors. But in nature the categories animals must form are less discrete and more open-ended. They must escape from predators, but not a predator that always looks like a red square of a certain size to be discriminated from a safe animal that might look like a blue square all the time. Instead they must escape from predators which, even if the same species, normally look somewhat different from each other. Furthermore, an animal's view of a predator when it begins its escape typically differs from one occasion to the next; at one time it may see the head of a lion, the rest of the body being obscured by vegetation, and at another time it preceives some other part of the body. But to survive it must consider all these instances of a lion as "predator."

To investigate in the laboratory open-ended categories similar to those faced in nature, Herrnstein and his colleagues (1976) have conducted various discrimination experiments using pigeons. In the first of this series they set several pigeons to the task of discriminating people from anything that was not a person. To accomplish this, the pigeons were shown 1,200 slides, of which 600 had one or more persons somewhere in the picture while the other 600 did not. In the photo, the person was close to the camera or distant, shown in entirety or just in part. The people were located in various parts of the photo and were children or adults, men or women, in various poses and from various ethnic groups. Using operant conditioning techniques, the experimenters rewarded the pigeons with food for pecking at a button when pictures of people were shown, but did not give the pigeons food if they pecked when photographs with no people in them were shown. By pecking at different rates for each type of picture, the pigeons seemed to be indicating that they could form a category of people versus nonpeople.

In further studies with his colleagues (Herrnstein, Loveland, and Cable 1976) Herrnstein investigated whether pigeons could form other categories such as trees, water, or a particular person. The tree could be either all or some part of the tree; the water, either a drop, a puddle, a lake, or an ocean. Cerella (1979) required pigeons to discriminate oak leaves from other species of leaves. The pigeons could learn all these discriminations, even differentiating some rather tricky ones such as a tree from a stalk of celery. And the pigeons were able to discriminate using both slides familiar to them and new to them.

What do these experiments say to us about basic aspects underlying the pigeons' ability to form categories? Perhaps pigeons have certain natural categories that they are innately equipped to form, such as trees or water. They may learn to make some distinctions, for example, discriminating a certain person from others. Yet pigeons can also learn to discriminate the capital letter A presented in various kinds of typeset from the number "2" (Morgan, Fitch, Holman, and Lea 1976). If there are genetic predispositions for categories, they are at least not straightforward. Perhaps pigeons, and other species as well, have the ability to form many categories that pay off for them, in this case get them some food. Perhaps pigeons have a general tendency to categorize things associated with each other, and they can refine this tendency further and further as it pays off for them to refine it.

There is another very interesting wrinkle to pigeons' ability to form categories. Herrnstein and de Villiers (1980) set pigeons to the task of discriminating between 40 slides portraying fish in various orientations and 40 slides without fish. The same 80 slides were used repeatedly until the pigeons were quite accurate in their discrimination. Then the question was asked, is it necessary that the pigeons learn categories? Could they instead be learning about the individual pictures per se? To answer this, another group of pigeons was given, as the positive category, 20 slides depicting fish and 20 slides without fish. The negative category also had fish and nonfish pictures, but consisted of different photographs. The pigeons could solve this task too, though it took them more than twice as long as it did for the first group to learn the fish—no fish discrimination.

All these experiments seem to be indicating, at least to me, that pigeons can learn to form categories if it pays off for them to do that, or they can learn to identify individual pictures if it pays to do that. The consequences matter. And, at least in some instances, it may be easier to form categories than to remember many individual photographs.

Notice that I have been carefully using the word "category" or "class" and generally avoiding the term "concept." It may be that we want to distinguish the ability of humans and perhaps of some other species to form a concept from the ability to form a category. One difference between these two terms is that if an organism can form a category, that does not imply it has an ability to abstract the relevant features or family of features, some of which characterize each example. So even from these laboratory examples of basic abilities that must underlie thinking, we may find important differences between humans and other species.

Studies of self-awareness

Another component of an animal's thinking concerns whether it is aware of itself and its activities. Does an animal know what it is doing when it is doing it? How could we go about answering such a question? If we were investigating humans, we could just ask them, but we can't ask an animal, or can we? In effect haven't we been doing something of the sort in the long history of animal experiments that ask if an animal can make certain discriminations? Rather than having the animal talking to us per se, we use its behavior to reveal a capacity that may initially have nothing to do with that behavior. Recall, for instance, the work of Herrnstein and colleagues in which pigeons discriminated between people, trees, and water. We learned that pigeons were capable of making such distinctions by rewarding them for responding to those classes of stimuli and then noting differences in rates of responding that developed.

Several experimental psychologists, Beninger, Kendall, and Vanderwolf (1974), have investigated whether rats can discriminate their own behaviors. In decades before them, others, even the behaviorist Hull, had done experiments to indicate that rats could discriminate between some internal states, in particular, hunger from thirst. Others showed that dogs and monkeys could discriminate between various sorts of visceral stimulation. Such findings suggest at least some animals other than humans may be aware of internal states. In the research done by Beninger and his colleagues, rats were trained to press a lever for a bit of sweet condensed milk. Once the rats had accomplished this task, the experimenters became more stringent and rewarded the rats only for lever pressing after a buzzer sounded. Actually, there were four levers present and gradually the rat was reinforced only for pressing a specific lever when it was doing a specific behavior. Each lever was associated with a different behavior that the rat tended to make in the situation—face washing, walking, rearing, and immobility. Thus, for example, if the buzzer sounded when the rat was rearing up, he had to press lever 4 in order to get a reward, and so on.

The rats could do this. Each of four rats responded by pressing most levers correctly 70 percent to almost 90 percent of the time. Some rats were occasionally more or less accurate than this. Over 99 percent of their responses were considerably better than chance.

The experimenters investigated other possible interpretations of the rats' performance. For instance, if a rat tended to perform each behavior type in a different location, it could be learning to use location, rather than its behavior, as a cue to appropriate lever pressing. However, location was not precisely correlated with the animals' activity. These experiments support an interpretation that the rat seems to have some sense of what it is doing, but whether it is by the sensory stimulation from its activities or a more global notion such as "I am now rearing up," we don't know. But animals, at least rats, seem to be able to discriminate between their behaviors. That seems an eminently useful ability to have. Recognize, however, that even these experiments are not ironclad proofs of awareness. It *is* still possible to argue that the rat is unconsciously associating some bodily sensation with reward.

We have just discussed whether an animal could be aware of its own behavior. Can we try something yet more tricky? Could an animal be aware of itself? Some aspects of the problem of self-awareness have been tackled by Gallup, who has defined self-awareness as the capacity to become an object of one's own attention. He noticed that chimpanzees, when given access to a mirror, initially react to the mirror image as though it were another conspecific. They may grimace toward it and do a number of "other-directed" or social responses. As time goes on, chimpanzees decrease responding socially to the reflection and, by about the third day, self-directed behaviors appear, such as touching parts of the body while looking at the mirror. In particular, chimpanzees seem very interested in touching body parts that are not visible without the aid of a mirror. Young children also react similarly to mirrors, and when about 10 months old appear capable of recognizing themselves in the mirror. We use the words "recognizing themselves in the mirror" fairly easily with human children, but what of the chimpanzees?

To make a more convincing demonstration that the chimpanzees were indeed knowingly, not randomly, touching parts of themselves that they saw reflected in the mirror, Gallup (1977) devised a formal experiment. After the chimpanzees had at least 10 days of experience with mirrors, he anesthetized and applied red dye marks to them. The dye, as far as could be determined, was free from tactile and olfactory cues. The marks were placed on the eyebrow ridge and at the top half of the opposite ear. When the chimpanzees recovered, they were again shown a mirror. Time spent viewing themselves suddenly increased and the animals touched the dyed spots numerous times, often visually inspecting and smelling the finger that had touched the marks, suggesting the chimpanzees expected something to be there. As a further control, the chimpanzees who had not been exposed to mirrors were also anesthetized and marked with the red dye. However, when these chimpanzees were placed before a mirror they did not touch the marks, but instead reacted to the mirror socially much as any chimpanzee does the first time it sees its reflection.

Can other species do these mirror reflection tasks as well? Many species of monkeys that had extensive experience with mirrors never progressed beyond other-directed behaviors, and when experimentally marked with dye, did not touch those marks. Various species of birds, including chickens, also did not progress beyond other-directed behaviors, and neither did socially isolated chimpanzees, those that had been reared from a very early age without social companions. Among the other great apes besides chimpanzees, orangutans exhibit self-recognition of mirror reflections, but gorillas do not (Suarez and Gallup 1981). When tested in the dye experiments gorillas do not increase looking at or touching the marked areas. Yet a gorilla does pay attention to a mark placed directly on its wrist that does not require a mirror to observe. This is a bit strange; do the results mean that, for reasons unclear, the gorilla cannot recognize itself? There is some opposing evidence, namely that the gorilla Koko, used by Francine Patterson in language training experiments, is reported to play with mirrors, applying gray make-up or poking various parts of her face while looking in a mirror (Patterson and Linden 1981). The difficulties could be motivational; perhaps the gorilla in the Gallup experiments simply could not be bothered to use the mirror to poke at marks on itself.

Further, another surprise. Despite the fact that various chickens do not make self-directed behaviors using a mirror, pigeons have been trained to peck at spots on themselves that they can see only in a mirror. In this experiment Epstein, Lanza, and the renowned operant conditioner B. F. Skinner (1981) trained pigeons to peck at a dot on their bodies that was hidden underneath a bib. A pigeon could see the dot only by raising its head high so that the bib held around its neck was lifted off the dot that was then visible in the mirror. Must we conclude that pigeons are self-aware while monkeys and gorillas are not? It certainly is possible that pigeons could be self-aware, and I don't mean to negate that possibility. However, I want to emphasize that the researchers did very exacting training of the pigeons over a long period of time. The pigeons were carefully trained to peck at blue spots as they appeared first on the front wall; then gradually the spots were moved to the rear walls and finally the spots were present only while the pigeon was looking in the mirror. After precise shaping of behavior, the pigeons were able to accomplish the task. This is markedly different from the situation faced by the chimpanzees and even gorillas. In the Gallup investigations, the animals were simply given mirror experience, that is, put in a cage with a reflecting surface. After they were anesthetized and marked, their spontaneous responses were noted. There was no specific training of the skill to touch marks on themselves. Furthermore, the operant conditioners who trained the pigeons had a long, successful history of pigeon training. It is not an easily replicable experiment and others who have tried it have gotten mixed results (Gelhard, Wohlman, and Thompson 1982).

What have we learned, then, from these experiments into the nature of self-awareness? For one, depending upon the kind of interpretations we want to draw, we must be wary of looking at the behavior an animal exhibits without knowing about the history of that behavior. Was it very specifically and carefully trained, as with the pigeons in the Skinner experiment, or indeed as with apes in the various ape language projects? If so, we may be talking about quite different capacities from those that can arise spontaneously from the animal. Second, what is it about self-awareness that has been indicated from these experiments? I think we must clearly separate a sense of self as mind from self as body. The self as mind includes the self that we are today, yesterday, and tomorrow—the self that has a certain style of thinking, or has a personality. These aspects of self have not been investigated in these experiments. Instead, aspects of self as an object, as a body, have been studied, and that is indeed interesting as well.

Studies suggesting attribution of mental states and intentional deception

Let us go a step further and ask whether an animal can attribute one aspect of mind, mental states, to another. Mental states that we commonly attribute to each other are "belief," "hope," "intention," "wanting," "liking," and so on. In a series of experiments Premack has attempted to investigate whether chimpanzees might be able to attribute mental states to humans. To do so, Premack and Woodruff (1978a) showed a chimpanzee a videotape of a human trying to reach an unattainable item such as bananas, or engaged in other more complicated tasks, and then offered the chimpanzee two photographs to choose between. The ape, of course, had previously been taught to play these choose-a-photograph games, for which it would get some reward. The photos the chimpanzees were given depicted either correct solutions to the problem, namely two boxes piled on each other if the bananas were placed out of reach overhead, or simply two boxes next to each other. Or the incorrect photograph could have objects that might be found in the environment but have little to do with the correct solution. In order to solve this problem, the chimpanzee presumably had to infer from the human's behaviors that the man wanted to reach the bananas. The chimpanzee then had to interpret the photographs as possible solutions for the human and choose the correct solution. Of course there are other reasonable interpretations we can place on this situation. It is, however, very difficult not to conclude that a chimpanzee must impute the intention that the human *wants* to reach the bananas. Otherwise the ape has merely observed a video of a human jumping up and down, arms outstretched, and a bunch of bananas overhead, all of which bear no obvious relationship to a photograph of two boxes piled atop each other. It is critical to such an interpretation that the ape has not watched a human engage in these behaviors before, for the test would then merely reflect the ape's ability to remember behavioral sequences. In the least, the experiments are an ingenious attempt to experimentally investigate the attribution of mental states. Premack and Woodruff conducted other experiments somewhat too complicated to discuss here, but interested readers may consult the December 1978 issue of *Behavioral and Brain Sciences* for these and other studies of animal cognition. The journal offers commentary from peers in various disciplines such as biology, psychology, philosophy, and anthropology to provide diverse viewpoints.

Another fascinating area of animal behavior that might successfully be used to reveal animal thinking and intention is deception.

Field studies have noted deceptive interactions, particularly among primates. A compelling case with langur monkeys has been captured on film by the researcher Sarah Blaffer Hrdy. In one episode a langur kidnapped another's baby. The sequence begins with a female approaching the vicinity of a mother and infant from a different langur troop. The approaching female was clearly looking everywhere except toward the baby. When the female was quite close to the baby, she suddenly dashed up, grabbed the infant, and ran off. In this case the deception involved an inhibition of eye gaze, in fact, the directions of eye gaze were completely inappropriate to the monkey's goal. Viewing such an episode, it is very difficult not to interpret the interaction as the female intending to kidnap the infant, and intending not to let the mother know that.

It would be very useful to have a formal system by which to evaluate the complexity of a deceptive interaction and to guide us in gathering evidence about attributing intention to an organism. There already exists a useful formulation that can be applied to apparently deceptive interactions and other forms of communication in order to better understand the degree of mental complexity involved. The formulation is known as intentional analysis and was developed separately by Daniel C. Dennett (1978a, 1983) and Jonathan Bennett (1976, 1978), and derived from work of other philosophers such as Grice. Colin Beer (this volume) discusses intentional analysis and its application to reciprocal altruism, so I will be brief in my discussion. First, when philosophers use the term "intentional," they do *not* typically mean "intending to" or "on purpose." Examples of intentional terms that can be attributed to an organism or anything else to which we might want to attempt to attribute them are "believes," "desires," "knows," "wants," "thinks that," and so forth. Very simplistically and loosely stated, they are terms we might colloquially call mental terms. We can describe the behavior and interactions of an animal according to different levels of intentionality. Depending upon the evidence we have available and the experiments we can do, we may be able to determine which level is the most accurate for describing that organism doing that particular sort of behavior. As an example, the zero order of intentionality is what is typically used in acceptable scientific description, that is, one that does not make use of mental terms. Some examples might be organism A locomotes or organism A makes certain gestures under specifiable stimulus conditions. A first order intentional statement might be A *knows* that the nest is under the tree, or A *wants* to find a mate. In the second order of intentional statement we would be dealing with the mental states of two entities, such as organism A believes that organism B believes that X is true. Or the ape believes that his caretaker believes that he broke the plate. In the third order we would say A wants B to believe that A wants or believes X. For example, the ape wants the trainer to think that the ape knows where the food is hidden. With intentional analysis, we can look at some apparently deceptive situations to understand what assumptions we are making about the mental states and beliefs of the organisms involved. Dennett has already applied intentional analysis to some aspects of the ape language work (1978b) and to other ethological work (1983).

With these ideas in mind, I shall now report just a few cases of deception. Kummer (1982) has made several observations of monkeys, specifically hamadryas baboons, engaged in apparently deceptive interactions. Hamadryas baboons live in harems with a single male leader and several of his females. He attempts to restrict the females to copulating only with him. On one occasion (mentioned by Beer in this volume), a female baboon in oestrus, that is, receptive for copulation, moved away from the adult male leader and repeatedly copulated behind a rock with a juvenile male. Their position behind the rock hid them from the adult male, although the female did look out at the leader. In fact, at times she approached the leader and presented herself to him, which is a reassuring gesture, and then left him again to copulate once more with the juvenile. On another occasion, a female spent almost a half hour edging herself behind a rock where she could groom a juvenile male in a position such that the onlooking male leader could see only the top of her head and her back, thereby missing her attentions to the juvenile. By an intentional analysis we could say that female did not want the harem leader to know about her activities.

Premack and Woodruff (1978b; Woodruff and Premack 1979; discussed in Ristau and Robbins 1982) used a laboratory situation in order to investigate intentional deceptive communication. The scenario was in some ways analogous to Menzel's (1974) work with chimpanzees. Woodruff and Premack explored the chimpanzee's ability to deceive and resist being deceived. In the experiment, a human and a chimpanzee communicated about the location of food hidden in one of two containers. The chimpanzee had previously seen the food being hidden and therefore knew where it was. A trainer then entered, either the "good guy" trainer who shared the food with the chimpanzee if the chimp found it on its first guess, or a "bad guy" trainer who kept all the food for himself. In the second phase it was the human who knew where the food was hidden. The "good guy" human gave various nonverbal signals indicating where the food was actually hidden, while the "bad guy" purposefully misled the chimpanzee. The chimpanzees dealt ably with the humans. In the cooperative situation, the chimpanzee succeeded in producing and comprehending behavioral signals about the food's location. However, when the human and chimpanzee competed for the food and the human kept all the food for himself, the chimpanzee learned to withhold information about the location of food, and to mislead the competitive human. The chimpanzee even spontaneously began to direct the "bad guy" to the wrong container by pointing, which is particularly interesting because chimpanzees have not been observed to point in the wild. Finally, when the chimpanzee was the recipient of cues from the "bad guy," it learned to discount the "bad guy's" misleading behavioral cues.

If we wanted to suggest a fairly high-order intentional analysis of this situation, the third order, we might say that the chimpanzee "wants the bad guy to think that it knows where the food is." But there is controversy as to whether this is deception. Some might say that after a considerable amount of training, Sadie, the chimpanzee, has simply managed to learn what behaviors to use to get a reward, namely food. There is something a bit odd about the situation in that the "bad guy" trainer does not ever discount the chimpanzee's misleading signals. And why doesn't the "bad guy" simply take the second container of food? But the experiment is, nevertheless, an important and insightful attempt to study intentional behavior in an animal.

I have become interested in the problem of "injury-feigning" birds (Ristau 1983a,b,c). There are a number of ground-nesting birds which, when they have a nest or young, will, in the presence of an intruder or predator, perform very awkward wing flapping that has come to be known as broken-wing displays or "injury feigning." The birds engage in other antipredator behavior as well. Of the various behaviors, the broken-wing displays have received particular attention because they are so dramatic and one is easily convinced that the bird is maimed, unable to fly, and probably easily caught (figure 2). But after displaying, sometimes for many minutes and hundreds of feet, the bird suddenly and agilely flies away.

Figure 2 The broken-wing display of a plover involves very awkward wing flapping that resembles an actual injury. Such behavior draws the attention of a potential predator away from the nest. (Illustration by Richard Swartz.)

Many anecdotes suggest the adult birds are leading the intruder away from the nest or young by this display, but this interpretation is not completely accepted. The display has indeed been described as a "hysterical" response made by a bird convulsed between parental drives to go to the nest and young, and conflicting tendencies to attack or flee from the intruder. Some consider the display to be a reflex made haphazardly in any random direction when the bird is in the appropriate hormonal condition and in the presence of a moving ground predator or object.

But could the bird be purposely leading intruders away? In a first order intentional analysis, we would say "the bird wants to lead the intruder away from nest or young" or "the bird wants the intruder to follow it." This is what we are studying. And if we find evidence for first order intentionality, we will attempt to explore the second order, such as "the bird wants the intruder to think the eggs or young are not located where they really are."

How are we doing this? With both verbal descriptions and sometimes videotape as well, we can conduct experiments in the field in which intruders walk in the vicinity of the nest or young and then approach the offspring. If the parent bird is trying to lead an intruder away, we can predict that (1) it should move while displaying in a direction that would alter the intruder's path away from the nests or young, (2) it should monitor the intruder's behavior, and (3) it should adjust its behavior in accordance with the intruder's behavior. From our work with piping plovers, we already have quantitative data supporting the first and third predictions (Ristau 1983a, 1984) and qualitative data for the second. We have also learned from our field experiments that the plovers can quickly learn to discriminate between human intruders who have walked "dangerously" close to their nest, and therefore may pose a threat to their eggs, and intruders who are "safe." We are continuing and expanding our investigations, and we hope to learn more about the intentionality of the injury-feigning birds.

Conclusion

Do animals think? We as scientists must continue to explore that issue, recognizing that various mental states an animal might experience are likely to be different from ours. We can use mental terms with animals to help us pose questions and perform experiments in the laboratory and in the field that may not even be initiated without hypothesizing such mental states. Then we can determine with more precision, making judicious but not excessive comparisons to human abilities, what manner of abilities can be included as animal thinking. To permit the use of mental terms in the description of animal and human behavior will, I think, allow us in the end to use more parsimonious explanations and integrate broader areas of behavior.

Acknowledgment

I would like to express my great debt to and appreciation for Donald R. Griffin, with whom I work, who is indeed a contemporary pioneer in the field of animal thinking. His ideas have influenced mine, though neither of us completely agrees with the other, which allows for enjoyable arguments.

Select Bibliography

Beninger, R. J., S. R. Kendall, and C. H. Vanderwolf. 1974. The ability of rats to discriminate their own behaviors. *Canadian Journal of Psychology* 28:79–91.

Bennett, J. 1976. *Linguistic behavior.* London: Cambridge University Press.

———. 1978. Some remarks about concepts. Cognition and consciousness in nonhuman species: Open peer commentary. *Behavioral and Brain Sciences* 4:557–60.

Cerella, J. 1979. Visual classes and natural categories in the pigeon. *Journal of Experimental Psychology: Human Perception and Performance* 5:68–77.

Darwin, C. 1859. *The origin of the species.* Reprinted 1968. Harmondsworth: Penguin Book.

———. 1872. *The expression of the emotions in man and the animals.* Reprinted 1965. Chicago: University of Chicago Press.

Dennett, D. C. 1978a. *Brainstorms.* Montgomery, Vt.: Bradford.

———. 1978b. Beliefs about beliefs: Open peer commentary. *Behavioral and Brain Sciences* 4:568–70.

———. 1983. Intentional systems in cognitive ethology: The "Panglossian paradigm" defended. *Behavioral and Brain Sciences* 6:343–90.

Epstein, R., R. D. Lanza, and B. F. Skinner. 1981. 'Self-awareness' in the pigeon. *Science* 212:695–96.

Gallup, G. G., Jr. 1977. Self-recognition in primates: A comparative approach to the bidirectional properties of consciousness. *American Psychologist* 32:329–38.

Gelhard, B. S., S. H. Wohlman, and R. Thompson. 1982. Self-awareness in pigeons—a second look. Paper presented at Animal Behavior Society meetings. Boston, October 1982.

Griffin, D. R. 1981. *The question of animal awareness*. 2d ed. New York: Rockefeller University Press.

Griffin, D. R., ed. 1982. *Animal mind—Human mind*. Berlin: Springer-Verlag.

———. 1984. *Animal thinking*. Cambridge: Harvard University Press.

Herrnstein, R. J., and P. A. de Villiers. 1980. Fish as a natural category for people and pigeons. In *The psychology of learning and motivation*, ed. G. H. Bower. Vol. 14. New York: Academic Press.

Herrnstein, R. J., D. H. Loveland, and C. Cable. 1976. Natural concepts in pigeons. *Journal of Experimental Psychology: Animal Behavior Processes* 2:285–302.

Humphrey, N. K. 1980. Nature's psychologists. In *Consciousness and the physical world*, ed. B. Josephson and B. S. Ramachandra, pp. 57–80. New York: Pergamon.

Jolly, A. 1966. Lemur social behavior and primate intelligence. *Science* 153:501–6.

Kamil, A. D. 1978. Systematic foraging by a nectar-feeding bird, the Amakihi (*Loxops virens*). *Journal of Comparative and Physiological Psychology* 92:388–96.

Knudsen, E. I., and M. Konishi. 1978. A neural map of auditory space in the owl. *Science* 200:795–97.

Kummer, H. 1982. Social knowledge in free-ranging primates. In *Animal mind—Human mind*, ed. D. R. Griffin, pp. 113–30. Berlin: Springer-Verlag.

Menzel, E. W. 1974. A group of young chimpanzees in a one-acre field. In *Behavior of nonhuman primates*, ed. A. M. Schrier and F. Stollnitz. Vol. 5, pp. 83–153. New York: Academic Press.

Morgan, C. L. 1894. *An introduction to comparative psychology*. London: Walter Scott.

Morgan, M. J., M. D. Fitch, J. G. Holman, and S. E. G. Lea. 1976. Pigeons learn the concept of an "A." *Perception* 5:57–66.

Nagel, T. 1974. What is it like to be a bat? *Philosophical Review* 83:435–50.

Olton, D. S. 1978. Characteristics of spatial memory. In *Cognitive processes in animal behavior*, ed. S. H. Hulse, H. Fowler, and W. K. Honing, pp. 341–74. Hillsdale, N.J.: Lawrence, Erlbaum Assoc., Inc.

Patterson, F. G., and E. Linden. 1981. *The education of Koko*. New York: Holt.

Premack, D., and G. Woodruff. 1978a. Does the chimpanzee have a theory of mind. *Behavioral and Brain Sciences* 4:515–26.

———. 1978b. Cognition and consciousness in nonhuman species: Authors' responses. *Behavioral and Brain Sciences* 4:616–29.

Pylyshyn, Z. 1978. Imagery and artificial intelligence. In *Perception and cognition: Issues in the foundations of psychology*, ed. C. W. Savage. Minnesota Studies in the Philosophy of Science, vol. 9. University of Minnesota Press. Reprinted in *Readings in philosophy of psychology*, ed. N. Block. 1980. New York: Harper & Row.

Ristau, C. A. 1983a. Do "injury-feigning" birds lead intruders away from nest and young? Paper read at the Animal Behavior Society meetings, Bucknell, Pa., June 1983.

———. 1983b. Intentionalist plovers, or just dumb birds: Commentary on D. C. Dennett. *Behavioral and Brain Sciences* 3:373–75.

———. 1983c. Language, cognition, and awareness in animals? In *The use of animals in biomedical research*, ed. J. Sechzer. New York Academy of Sciences.

———. 1984. Intentional behavior by "injury-feigning" birds? Paper read at the American Psychological Association meetings, Toronto, August 1984.

Ristau, C. A., and D. Robbins. 1982. Language in the great apes: A critical review. In *Advances in the study of behavior*, ed. J. Rosenblatt, R. A. Hinde, C. Beer, M.-C. Busnel, Vol. 12, pp. 141–255. New York: Academic Press.

Romanes, G. J. 1882. *Animal intelligence*. London: Kegan Paul, Trench.

———. 1884. *Mental evolution in animals*. New York: Appleton-Century-Crofts.

Suarez, S. D., and G. G. Gallup, Jr. 1981. Self-recognition in chimpanzees and orangutans, but not gorillas. *Journal of Human Evolution* 10:175–88.

Woodruff, G., and D. Premack. 1979. Intentional communication in the chimpanzee: The development of deception. *Cognition* 7:333–62.

From Aristotle to Descartes: Making Animals Anthropomorphic

Stephen J. Vicchio

Anthropomorphism is, quite simply, the tendency to project or ascribe human qualities to nonhuman animals, events, or objects (Agassi 1964). We use it when we speak of the wind blowing a ship off course, or the sea tossing about a helpless craft. William Gilbert was guilty of it in 1600 when he described the action of magnetic attraction and repulsion as love and hate. James Clerk Maxwell (1873) was equally culpable in his *Electricity and Magnetism*, in which he compared Faraday's tubes of force to human muscles. But perhaps the most pervasive and consistent subjects of anthropomorphism are the members of the animal kingdom.

The history of humankind's attitudes toward animals is very complicated. In this paper I will review the historical tendency to make animals anthropomorphic in Western culture, and present examples of several different styles of stereotyping and anthropomorphism that seem to have survived to the present day. Let us begin, however, with some preliminary observations.

When we make use of anthropomorphism to explain animal behavior we have already implicitly assumed that we know human nature sufficiently well to find ourselves in animals. We assume, more or less, that an analogy that moves from the known (humans) to the not so well known (animals) can be made. Ironically, this may present us with a problem from the start. If we put the problem in a positive way, it is that we may know as much about the animals as we do about ourselves; to put it in what some thinkers feel is the more realistic negative, we may know just as little about ourselves as we do about the animals.

Another consideration that may be helpful to keep in mind involves what is often in philosophical circles called the genetic fallacy. It essentially argues that if you know the origin of an idea, you know that idea's veridical worth. But we must keep in mind that whenever we make an anthorpomorphic assertion about a given animal, it may be true or false. It is never a sufficient criticism of an idea to show simply whence that idea has come. Some anthropomorphic notions of animals are known to be false, but they are not false simply by virtue of the fact that they happen to be anthropomorphic. To say that some animals behave like humans in some ways may be anthropomorphic, but the claims may still be true. However, with all of this said, the greater danger usually lies in assuming uncritically that anthropomorphic attributes are true of animals. We will return to this point. It would be more profitable now, however, to turn to some general historical comments about our relationship to the animal world.

A Brief History of Anthropomorphism and Attitudes toward Animals

The history of ideas about the wisdom, stupidity, cleverness, kindness, prudence, faithfulness, slyness, and clairvoyance of animals is long and complicated. The number of assertions that can be made with any certainty about humans' earliest attitudes toward animals, however, is very small. It does appear nevertheless, that early perceptions of animals were an admixture of fear, love, distrust, curiosity and, above all, awe. We have a small quantity of visual evidence from the Stone Age that may help us in interpreting these early attitudes. Perhaps the earliest records are painted on the walls of caves in Lascaux and Altamira (figure 1).

What made these people, who apparently lived by hunting, cover their walls with these vivid images? Prehistorians have given a variety of answers to this complicated question. One suggestion is that it gave these hunters power over their prey. Through the use of the primitive's law of similarity, the cave dwellers produced images of animals, and having control over these images, had control over the beasts as well.

This interpretation is quite unsatisfactory, principally because it does not fit the drawings. For one thing, the people depicted in these cave drawings are always much smaller and less detailed than

Figure 1 The aurochs, a large ancestor of modern cattle, was depicted on cave walls in Europe in late prehistoric times. (Illustration by Margie Gibson.)

the animals represented. If they had desire for control, why not make the humans depicted more formidable? Kenneth Clark (1977) believes these drawings are portraits of admiration. As he puts it: "They say, 'this is what we want to be like'" (p. 14). This is confirmed, Clark suggests, by man's next step in his relationship to animals: the choice of animals as sacred symbols—the development of totemism.

Totemism has probably existed all over the world, but it appears to have been strongest and most complex in Africa. It may well be that in ancient Egypt we see the first example of totemism turned into a full-scale religious world view. So strong were the totemic practices of these earlier proto-Egyptians that we see, from the earliest of recorded Egyptian civilization, art that attempts to integrate people and animals. The Egyptians believed the human body to be a perfect anatomical shape. They placed the heads of the former totemic animals on the shoulders of these perfect human forms, and the result was the gods (figure 2). (Later, the Greeks would perform the reverse of this process. They took the upper body of the human, complete with the face—for the Greeks, the mirror of the soul—and grafted it to the powerful haunches of an animal, thus creating centaurs and harpies—see figure 3.) Of all the ancient Egyptian deities, Horus, the hawk-headed god, seems to have been the most often depicted. The Horus relief in the Louvre is a magnificently regal surviving example. The other sacred figures of the Egyptians could be ranked in a hierarchy of sanctity with Hathor, the cow, favored among certain pharaohs, the ram being sacred to Amun, and Ibis and Toth, the ape, each being the totemic choice in some locales.

In ancient Sumeria, animals were most often depicted as strong

Figure 2 The Egyptians portrayed many of their gods as having animal heads and human bodies. (Illustration by Vichai Malikul.)

and ferocious. Kinship with the animals was superseded by awe at their power and strength. Lions, bulls, and snakes were most often the sacred choices of cultures of the ancient Middle East. Their strength, potency, and in the case of the snake, its phallic shape, made them obvious symbols of fecundity and power in these war-prone kingdoms. The prominent role played by the snake in the mythology of the Canaanites perhaps gives us a clue as to why the ancient Jews converted this image in the ninth century B.C. from a symbol of wisdom to, in Genesis, a symbol of cunning and deceit.

Fables and myths depicting the attributes of animals can be found in early Egyptian, Sumerian, and Indian cultures. The Sanskrit book of parables, the *Panchatantra*, as well as the Pali Buddhist text, the *Jakata*, are full of such tales, and many of these same stories reappear in Aesop's *Fables*, and again in *Reynard the Fox*, a popular French text written by Jean de la Fontaine in the thirteenth century. These stories are usually very amusing and never without morally instructive value, and it is doubtful that they were ever meant to be taken as the literal truth about the attributes of animals. The same can be said for the talking beasts in the ancient Greek "Battle of the Frogs and Mice," as well as those animals depicted in Aristophanes' comedies. All of these creations seem to be more concerned with the parodying of human folly and lack of virtue than they are with telling us something important about the nature of animals.

There are, however, one or two of Aesop's *Fables* (1873 edition) that may in fact carry us back to an earlier age when our perceptions of animals may have been quite different. In many cultures in the ancient world there appears to have been a belief in a primeval period of total peace and harmony in nature, when the lion lay down with the lamb. In the Judeo-Christian tradition, we find this story in the myths about the Garden of Eden. Remnants of this belief can also be found in the later myths of Orpheus. There is in the Florentine Bargello a leaf on an ivory diptych that shows Orpheus accompanied by the animals; he sits apart from them and smiles down in their direction with a dreamy expression on his face. It dates from the fourth century A.D., when representations of Orpheus were still relatively common. The best indication of this earlier belief in a peaceful state of nature, however, can be found in Aesop's "The Lion Kingdom":

Figure 3 Centaurs were a race fabled to be half man and half horse living in the mountains of northern Greece. Centaurs were considered aggressive to humans. (Illustration by Richard Swartz.)

Making Animals Anthropomorphic   191

The beasts of the field and forest had a lion as their king. He was neither wrathful, cruel nor tyrannical, but just and gentle as a king could be. He made during his reign a proclamation for a general assembly of all the birds and beasts, and drew up provisions for a universal league in which the wolf and the lamb, the panther and the kid, the tiger and the stag, the dog and the hare should live together in perfect peace and amity. The hare said "Oh, how I have longed to see this day, in which the weak shall take their place with impunity by the side of the strong" (p. 56).

It is doubtful, of course, that this pre-Fall world ever existed outside the imagination, but echoes of it can be found in various references to the belief that animals formerly talked with human voices, a kind of reverse Dr. Doolittle. By the time of the Homeric Greeks, however, few traces can be found. In Homer's *Iliad*, Clytemnestra's comments about Cassandra, when the latter has been transported to Agamemnon's palace, may point to this prior belief in talking animals:

Unless like swallows she does use
Some strange barbarian tongue from over sea
By words must bring persuasion to her.

The humanizing spirit of the Homeric Greeks centered more on animal potency than power and ferocity. The bull became a perfect embodiment of Zeus, eloping with the not unwilling Europa. Zeus, who also borrowed the swan's plumage to woo Leda in the land of Laconia, was also often depicted as a brave eagle, while proud peacocks were associated with Hera, wise owls with Athena, and loving doves with Aphrodite.

What relation these ideas about embodying the gods may have had with earlier notions about transmigration of the soul is a topic that unfortunately goes beyond the scope of this paper. It suffices to say that the possibility for anthropomorphism becomes much greater when there is a concomitant belief in the reincarnation of souls from human to animal bodies. It is also clear that a number of prominent ancient Greeks believed quite positively in transmigration. Pythagoras, for example, suggested that he had been one of the Trojan heroes, whose shield he knew at a glance in the temple of Juno, where it had been hung.

By the time of Periclean Athens, Socrates, perhaps borrowing from this set of much earlier beliefs, suggested that imperfect earth-bound spirits might be reincorporated in animals whose conventionally ascribed characteristics correspond with their own moral natures: unjust, tyrannical men would become wolves or hawks, while good, commonplace, industrious folk were destined to become ants, bees, and wasps. Socrates' friends appear to be concerned when they do not find him in a spirit of depression a few days before his execution. The philosopher asks his visitors if it appears to them that he is inferior in divination to the swan who, when it perceives that it must die, is given to lyrical song. He suggests that mankind falsely believes that the bird is singing because it is grieving and fearful. The real reason, says Socrates, is that the bird has foreknowledge of the bliss of Hades and is expressing it in the joy of song the day before it is to die.

Aristotle's perspective on animals seems far more naturalistic than those of Socrates and Plato. Although by modern standards his point of view is a curious combination of natural history, empirical observation, and the myths and symbols of an earlier age, Aristotle's *History of Animals* was still consulted as a zoological text well into the sixteenth century. It was originally written at the request of Alexander the Great who, the Greek historians inform us, had been guided by crows sent from the gods to show him the way across the Libyan desert.

In one section of his works on animals, Aristotle waxed eloquent about the moral qualities and intelligence of the beasts. Men and mules are usually tame. The ox is gentle, the boar is violent. The serpent is crafty, the lion is noble and generous. Bears carry off their cubs at a sign of danger. Dolphins are to be commended for their extraordinary love of their young. He also suggested that deer can be captured by singing to them, bees sting human beings because the insects dislike bad smells, and that females of any species are less likely to help males in distress than vice versa. Of all the animals Aristotle discussed, the lion and the elephant are by far his favorites. The lion is described as gentle when not hungry, and never suspicious. The elephant is assigned the palm of wisdom, for he is a creature abounding in intellect, kindness, and fine memory. Later Apollonius would add that at night elephants mourn over their lost liberty with peculiar wailing sounds. If a person approaches, they cease their crying out of respect for the humans. Aristotle also informed us that in keenness of senses, man is by far surpassed by the other animals. This remark was later endorsed by Thomas Aquinas, who suggested that if we had the dog's keen sense of hearing, we would probably make bad use of it anyway.

But Aristotle's most enduring legacy regarding animals was his suggestion that man alone can reason, though he did believe other animals could remember and learn. Later, this question of animal intelligence would be taken up by various other ancient thinkers. At the beginning of the Christian era, Plutarch wrote the dialogue *Grillus,* in which Odysseus is allowed to speak with some Greeks whom a witch has turned into various animals. In the text, a precocious pig argues rather forcefully for the wisdom and prudence of swine over men, and in fact, states his preference for life as a pig.

A little later in the Christian era the heretic Celsus takes up the question of animal intelligence. Unlike other early Christians, he favored the theory that animals possess souls. He also denied that reason belongs only to humans, and suggested very strongly that God created the universe for all the beasts, not just one select species. He posited the view that "only absurd pride can engender the thought that humans are in any way special." Men seem different from animals because they build cities, make laws, obey rulers, and so on. But he pointed out that ants and bees do all of these things. If someone looked down from the heights of heaven onto earth, it would be difficult to distinguish our actions from those of the bees.

Celsus argued that man mistakenly feels he is superior because he possesses a soul that makes him capable of having notions of the Divine. But what is more divine, he asked, than the ability to tell the future? Birds give abundant indication that they are soothsayers. If birds and other animals show us signs of the future, it proves that they have a closer relationship to God (Lovejoy and Boas 1935).

Plotinus, a Roman Neo-Platonic philosopher, also suggested that animals have souls and an intelligence like ours. Accepting this postulate as true, he concluded that it should be unlawful, under any circumstances, to kill or feed on animals. Although Neo-Platonism made many important contributions to the life of the early church, Plotinus's attitudes toward animals did not become a majority opinion (Lovejoy and Boas 1935). By the time of Augustine, the fourth century, few Christians were vegetarians.

One important change that occurred in the beginning of the Christian era was the ending of animal sacrifice. The symbolic lions and bulls of the Canaanites, Hittites, and Assyrians were replaced briefly by fish and the Christian lamb and sheep. The Lamb of God, who takes away the sins of the world, replaced the old sacrificial animals. Although the lamb and sheep became the primary symbols for Christ's salvific act, the older religious images of the bull, lion, and eagle were reintroduced into Christianity in the second century by Irenaeus. By the end of the second century A.D., these older symbols began to be used not as symbols of the sacrificed Saviour but rather as representations of the evangelists. Two centuries later, St. Jerome suggested, in his Commentary on Ezekiel, that the symbolic animals depicted in that book are precursors of John the eagle, Mark the lion, Luke the bull, and Matthew the man.

From Augustine to Thomas Aquinas, Neo-Platonist thinking and the Christianization of Aristotelian ideas forced the debate about animal intelligence to recede into the background. Since animals did not have the higher part of the soul, the rational element, St. Thomas assigned them a place in the great chain of being below those of angels and men. Ironically, however, the animals were often still held morally responsible for their transgressions. The benign St. Francis, for example, did not stretch out a finger to help a wretched and gluttonous robin who, having perched on the edge of a vase, had tumbled in and drowned. It has been reported that St. Dominic tore a sparrow to pieces because he believed it was sent by Satan to disturb his prayers. Friar Bartholomew, writing on the subject of dogs in his *Encyclopedia*, suggested that the canine propensity for burying bones is due to anxiety and greed. He also believed that dogs as a group are wrathful, malicious, indolent, deceitful, and lazy. He made this last observation by offering the opinion that when dogs are bitten on the ears by flies, they will not respond to the attack because they lack motivation.

Another indication of animals as moral agents is the number of cats who were regularly executed along with witches in many West European towns during the Middle Ages. In fact, numerous allusions to animal trials can be found in Western history as early as the ninth century, though they may have existed earlier. Most of what we know about these trials comes from the works of three Frenchmen, Michael Rousseau, Jean Vartier, and a canon lawyer, M. Chassenée, who defended the practice of executing animals in his *De excommunicatione animalum insectorum*. The chief victim of these animal trials was, by far, the pig (Barloy 1974):

In the year of grace 1348, or thereabouts, a pig was apprehended at Fontenay after it had eaten a child, one Etienne la Camus, and the pig was burned in the courtyard of the town hall of the said place.

A trial that began in September of 1370 took place after the death of the son of a Burgundian swine herder (Barloy 1974). The boy had allegedly been killed by three sows who, according to the court report, seemed to have feared an attack on one of their young. All members of the herd were arrested as accomplices, which was a serious matter for the pigs' owners, a group of nuns. The sisters argued that the three pigs alone were culpable and the rest of the herd should be set free.

Apparently the dispute became quite complicated, for the decision of the court was not rendered until September 1379. At that time a magistrate ordered the three guilty pigs and one young sow to be executed. There is no indication in the record of what the fourth pig might have done. At any rate, this may be the only case in human history of a sow being turned into a goat. One wonders, of course, if the original three guilty pigs were alive nine years later.

The fresco in the Church of the Holy Trinity at Palaise depicts the execution of a pig in 1386. The swine had been sentenced to death on charges of killing a child. To make it look more like a human being, the animal was dressed in a jacket and trousers, as well as white gloves for its front paws. The record points out the pig's execution drew quite a large crowd.

Perhaps an even more interesting animal civil suit took place in Savoy in 1587 (Barloy 1974). The accused was a certain fly. Two advocates were assigned their respective sides in arguing the case of the fly. The defense argued that the fly and his comrades had been blessed by God, given the right to feed on grass, and thus were in the right when they occupied a certain vineyard. The prosecution countered by pointing out that the Bible and common sense showed animals to be created for the utility of man. Hence, the fly and his companions did not have the right to cause man harm.

The judge offered a compromise, suggesting that a piece of land might be made available where the fly and his friends could dine unmolested. On June 28, 1587, the citizens of the town hastened to the square to ratify an agreement that handed over the tract of land to the flies. The fly's attorney, however, complained bitterly that the piece of land was dry and barren. The townspeople denied it. Unfortunately, the end of this story is not to be found in the archives of the town, St. Julien.

In all, there are records of at least 150 medieval and modern trials of various animals. All of these trials may have had their origins in a certain reading of Exodus 21 : 28–32, which states the an ox that gores a person to death should be stoned. There were also countless numbers of horses shot between the eighteenth and twentieth centuries for bucking and/or killing their riders, not to mention the rather bizarre case of the hanging of an elephant in Baltimore in the beginning of this century.

In the medieval period we also see the development of bestiaries—richly illustrated compendiums purporting to describe the nature and attributes of various animals. The source for these manuscripts is often unmentioned, though the authority of Physiologus is sometimes quoted (Carlill 1924). If Physiologus existed at all, he may be a figure of antiquity, but it is more likely he was an early medieval personage. Most of the information found in bestiaries can be traced to Pliny, to the Latin texts of Aristotle, and to various versions of Aesop that were very popular from the tenth to fourteenth centuries. Bestiaries, like their classical forerunners, were based on observations, but they were also laced with a good number of anthropomorphisms. Eventually, by the sixteenth century, bestiaries became an avenue for poking fun at ecclesiastical and political excesses, and were no longer, strictly speaking, about animals.

The age of the symbolic animal appears to end in the seventeenth century with the work of René Descartes. His major thesis concerning our topic is that animals are essentially automata—that they are little more than sophisticated machines. It appears he gleaned this idea from Gomez Pereira, a Spaniard. Apparently Descartes forgot to thank Señor Pereira, who sued the Frenchman for plagiarism.

We may get the sense of Descartes's view of animals from the following quote (cited in Cesaresco 1909, pp. 353–55):

Because the clock marks time and the bee makes honey, we are to consider the clock and the bee to be machines. Because they do one thing better than man and no other thing so well as man, we are to conclude that they have no mind, but that nature acts within them, holding her organs at her disposal.

Nor are we to think the way the ancients did, that animals can speak, though we do not know their language, for if that were so, they, having several organs related to ours, might easily communicate with us as with each other.

Another Frenchman, a M. Chanet, echoing his master Descartes, asserted that he would believe that beasts could think when they told him so.

The most powerful negative argument Descartes felt he had given concerning the ability of animals to think was that they clearly did not possess immaterial souls because they have no pineal gland. Descartes, of course, was wrong about their anatomy, but he also outlined a metaphysical proof, which essentially entailed the notion that reasoning involves the knowing of universals that are immaterial and unchanging. Only a nonmaterial, unchanging soul can know these universals, and so, in short, animals cannot think.

By the seventeenth century, animals had been demystified in popular art. With the sixteenth century had come a new and modern aesthetic appreciation of the animal kingdom. By the seventeenth century, the demythologization was nearly complete. It has often been remarked that Descartes's philosophy marks the beginning of the modern era. The same can be said about his attitude toward animals. It would, of course, be beyond the scope of this paper to suggest, even in an outline form, the variety of cultural views of animals that can be seen since that time. What I would like to suggest, however, is that there are at least six ubiquitous stereotypes and styles of making animals anthropomorphic that because of the historical precedents we have just examined, appear quite frequently in twentieth-century Western popular culture.

Styles of Stereotyping and Anthropomorphizing Animals in Western Culture

The full range and complexity of these stereotypes and anthropomorphisms defies neat and easy classification. What I wish to do is note the more popular and influential types. It should be kept in mind, however, that none of these perceptions operates in isolation. Indeed, many of our attitudes about individual animals are a complicated composite of several of these types. At any rate, the images I wish to discuss are animal as alien, animal as child, animal as moral paragon, animal as demon, animal as clairvoyant or intuitive genius, and animal as machine.

Animal as Alien

Cardinal Newman once remarked that we know even less about the animals than we do about the angels. Presumably, we do know how many hippopotamuses can dance on the head of pin, but his point, I think, is still well taken. Ludwig Wittgenstein (1958) points to this same conclusion when he writes, "If a lion could talk, we would not understand him" (p. 223). The encounter of human and animal is always a meeting of radically different kinds. It is not surprising, then, that humans have in the past regarded as alien any animal whose behavior or physiological characteristics strike us as strange. As is sometimes the case with human aliens, we tend to be distrustful of these animals. Indeed, our attitude toward them is often a curious mixture of fascination and distrust. If you have ever stared into the eyes of a cat at dusk you have some sense of this feeling of mystery and fascination. The bat may be even a better example of animal as alien. On the one hand, we seem to be fascinated by its sensory abilities; on the other, we worry that it will become entangled in our hair.

What is ironic about animals as aliens is that our fascination with them would appear to derive from their dissimilarity to humans, while the fear or suspicion would seem to come from the realization of crucial similarities. We know what humans can and will do to each other, and we sometimes project those fears onto certain animals.

Animal as Child

The tendency to view animals as children, like the institution of childhood itself, is a relatively modern development. Although we can trace the notion of animals as dependent children directly back to figures such as St. Francis, the real roots of the animal-as-child anthropomorphism are to be found in the Romantic ideals of the late eighteenth and nineteenth centuries.

In fact, the growth of the notion of animal as child may run concomitantly with the beginning of the humane movement. The founding of the Society for the Prevention of Cruelty to Animals in Great Britain in 1824, as well as the A.S.P.C.A. in this country in 1866 by Henry Bergh, are good places to look for the beginning of the animal-as-child notion.

Most of the earliest members of these societies appear to have believed that animals possess souls. Appeals for the reform of attitudes toward animals were more often than not based on a perception of animals as poor defenseless children. In the ensuing years, many factions of the humane movement have continued this paternalistic and somewhat sentimentalized outlook. The remarks of Hans Ruesch (1978, p. 45) illustrate this point:

> The desire to protect animals derives from better acquaintance with them, from the realization that they are sensitive and intelligent creatures, affectionate and seeking affection, powerless in a cruel and incomprehensible world, exposed to all the whims of the master species.

In contemporary America, the animal as dependent child has considerable appeal. Walt Disney has made us into a culture of cuteness. The emphasis on cute and cuddly animals predisposes children to relate to animals in an unrealistic and anthropomorphic way. In the animal-as-child perspective, the value of the animal consists in its docility, playfulness, and charm as a human companion. Animals that do not meet the test of large head, sloping forehead, and wide-set eyes may be written off as not being worthy of human attention. Perhaps the best example of animal as child is the cartoon character Snoopy, who looks very little like a dog. His head is more than twice the size of the rest of his body, an exaggeration of the large head of young, human children. Mickey Mouse is now 50 years old, but he has retained the appearance and the voice of a prepubescent human being (figure 4).

It is of some interest to note that the thoroughly competent, self-assured child rarely strikes us as cute. In *Escape from Childhood*, John Holt (1975) argues that cuteness is the handmaiden of subjugation. He suggests that there is a direct correlation between sentimental perceptions of children and cruelty toward them:

> The trouble with sentimentality, and the reason sentimentality leads to cruelty, is that it is abstract and unreal. We look at the lives, concerns and troubles of children as we might look at

Figure 4 Although Mickey Mouse was created 50 years ago, he is still depicted in a condition of neoteny—he retains the appearance and voice of a young human being. (Illustration provided by Walt Disney Productions.)

actors on a stage, a comedy as long as it does not become a nuisance. And so, since their feelings and their pain are neither real nor serious, any pain we cause them is not real either. In any conflict of interest with us, they must give way; only our needs are real (p. 85).

It is the unreality of the cute-child image that creates difficulties. Similarly, it is the falseness of the child-animal image that provides a distorted picture of the nature and needs of certain animals. Children raised in an environment where a large number of animals are seen as cute, cuddly, docile, and innocent are ill prepared to see animals as animals.

Animal as Moral Paragon

In this kind of anthropomorphism the animal is seen as embodying man's best moral attributes. Good examples of animals as moral paragon are Rin Tin Tin, My Friend Flicka, Flipper, Lobo, Silver (the

Making Animals Anthropomorphic   201

Lone Ranger's horse) and, of course, Lassie. Timmy's dog Lassie was always morally perfect. While various other characters had bouts of indiscretion, she was a model of morality.

In most cases of the animal as moral paragon, the animal usually retains the physical appeal of the animal as child, while at the same time it embodies a cluster of moral ideals that normal human beings often seem to lack. Medieval literature abounds with stories of dogs helping to solve crimes. Much earlier Pliny had also told stories of dogs who had spotted their masters' murderers and reported them to the proper authorities.

Ernest Thompson Seton's animal books are another prime example of this kind of anthropomorphic activity. In the preface to *Lives of the Hunted* (Seton 1901), he tells us that the aim of the book is "to emphasize our kinship with animals by showing that in them we can find virtues most admired in man." He informs us that "Lobo stands for dignity and love-constancy; Silverspot for sagacity; Redruff for obedience; Bingo for fidelity; Visen and Molly Cottontail for motherloves; Wahb for physical force; and the pacing mustang for love of liberty."

We can certainly perceive a precedent for this kind of anthropomorphism in Aristotle's *History of Animals,* where the ancient philosopher speaks of the elephant in the most laudatory of ethical terms. This point of view can also be found in Pliny, the Psalms of David, and the preachings of Jesus. Among the relics of Pompeii is a dog lying across the body of a child, whom some suggest the canine was trying to save from the deadly lava. In 1808, Lord Byron left less room for debate about his intentions when he constructed a monument to his deceased dog, Boatswain. It describes the dog as having possessed beauty without vanity, strength without insolence, courage without ferocity, and all the virtues of man without his vices.

But in critically evaluating the model of animal as moral paragon we must recognize the inherent possibility of a dark side to this picture. If these virtuous animals can do no wrong, there are also animals who are often seen as nothing short of demonic.

Animal as Demon

If the animal as moral paragon is seen as attractive and forthright, the animal as demon is usually considered by human standards to be ugly and devious. Rats, snakes, vultures, sharks, wolves, scorpions, spiders, and, from the old Tarzan movies, crocodiles, are probably the best examples. These animals are often seen as treacherous and devious predators. Sometimes the animal is elevated to the status of the demonic incarnate, for example black cats, the shark in the movie *Jaws*, and the various medieval depictions of the serpent. Because of this association, there is a long history in the West of the abuse and torture of certain animals thought to be incarnations of the Devil, or at the very least, in league with him.

One noncanonical gospel tells the tale of the young Jesus curing a boy possessed by the Demon. When the Devil leaps out of the boy's mouth, it is in the form of a mad dog. Cerberus, of course, was the bronze-colored, multiheaded dog who guarded the underworld in Greek mythology. In Norse mythology, the blood-splattered dog Garm had the same job.

In early Greek mythology, Mother Earth sent a deadly giant scorpion to kill handsome Oron, but Oron flees into the sea to avoid confrontation. In Egypt, the evil god Set transformed himself into a scorpion to kill Horus.

The image of the animal as demon is anthropomorphic because it is a projection of the dark side of humanity onto animals. The appeal of the animal-as-demon anthropomorphism may well derive from our inability or unwillingness to squarely face the destructive tendencies in ourselves. When we project evil onto animals, we find it anew and combat it as an external enemy. In the animal-as-demon concept we may have operating what Jung has called the "projection of the shadow self."

Animal as Clairvoyant or Intuitive Genius

As we have seen in the earlier historical accounts, there is a long tradition in Western culture of ascribing extraordinary wisdom, and at times omniscience, to certain animals. The ancients thought the elephant and owl possessed great wisdom. Pliny thought that various birds were capable of human wisdom. Horses abound in Shakespeare's plays, though he does not mention by name the most famous horse of his day, "Morocco," who was known to restore gloves to their rightful owners after the horse's master had whispered the identity of the persons into the animal's ear. Morocco was also said to be quite efficient at doings sums as well as calculating the number of pence in any silver coin.

In Dante's *Inferno*, the spirit of Guido Montefeltro, a political figure of Florence, informs us that he did not perform the deeds of a lion, but rather those of a fox. Dante places Montefeltro in Hell, but three centuries later, Machiavelli places squarely on the throne any ruler who can imitate the strength of the lion and the cunning of the fox. In modern times we find elements of the animal as intuitive genius in many of the Disney movies.

Sometimes the animal is not just brilliant but also clairvoyant. Countless stories appear in Western literature describing dogs and other animals who refused to set sail on particular ships that later went to their murky deaths on the high seas. Socrates' example of the clairvoyant swan apparently was a fairly widely held belief at the time. In much more recent times, a scientist in California has suggested we can detect the approach of an earthquake by ascertaining the number of cats missing in neighborhoods lying along volatile areas.

The origins of divination and totemism are certainly related to this variety of anthropomorphism. Sometimes this notion converges with the idea of the animal as moral paragon to produce, as in the case of Lassie, a trusted savior, or when combined with animal as demon, an almost invincible foe.

Animal as Machine

Finally, we have the image of animal as machine. The roots of this point of view go back at least to Aristotle, but they are perhaps more clearly discernible in René Descartes's notion of animal natures. A more immediate precedent, however, can be found in Western industrialization and organization. After the Industrial Revolution, as more and more of life began to be seen in a mechanical way, it was perhaps inevitable that we would begin to comprehend animals mechanistically.

Two of the clearest representations of this kind of stereotype can be seen in certain farm and laboratory animals. The breeding, maintenance, and feeding of these animals are often controlled by very strict standards. The metaphor for large commercial endeavors involving farm animals is "quality control." Laboratory animals are bred for specific utilitarian purposes; the end product is experimental data.

In all of this there is a tendency to sometimes reduce these animals to cogs in a grand production system. This tendency is not, of course, without its parallels in the relations among human beings. In that regard this stereotype may also be an anthropomorphism in that we have taken the metaphor of the workers on an assembly line, or the systems management approach to personal issues, and applied it in an anthropomorphic way to various farm and laboratory animals.

Concluding Comments: Seeing Animals as Animals

This brings us to consider seeing animals *as* animals. Given the variety of ways in which we have historically seen animals, how can we see animals the way they really are? What is the proper way to view them? I would suggest that the answer to these questions must begin with an admission: we must start by acknowledging the mystery of animal natures. Just as human beings are more than the sum of their parts, the same can probably be said for the animals. The "abyss of incomprehensibility" about other species may be narrowed by careful, methodical, and humane research, but it will not be entirely bridged. Dr. Doolittle does not exist, nor do the animals with whom he and St. Francis seemed so intent on talking. Much of the anthropomorphizing of animals, I suspect, comes from a lack of knowledge about them. Books such as this volume, and other zoological programs that are careful to avoid these stereotypic and anthropomorphic sentiments, are sorely needed.

A second suggestion for better seeing animals as animals is to become more informed in a general way, not just about the current research being done about animal behavior, but also about our collective history. By being cognizant of the development of certain historical attitudes toward animals, we may become more aware of contemporary stereotyping and anthropomorphizing of animals.

A third suggestion is that we must attempt to see all species, particularly those traditionally thought to be unattractive, as being intrinsically valuable, despite how well they do or do not suit us as companions. We must begin to see with a unified eye, rather than an exclusionary one. We must begin to look at all animals, as well as ourselves and the rest of nature, as a grand ecosystem in which each of the elements contributes to the completeness of the whole. When we have accomplished this task, we will have ceased using animals as a clouded mirror through which we unconsciously attempt to find ourselves.

Select Bibliography

Aelian. 1923. On the nature of animals, trans. L. Thorndike. In *History of magic*. Vol. 1. New York: Macmillan.

Aesop. 1873. *Aesop's fables*. Reprinted in H. Frank, *The animal kingdom*, trans. Heinrich Stein-Howel. Stuttgart.

Agassi, J. 1964. Analogies as generalizations. In *Philosophy of Science* vol. 34, no. 4, pp. 351–56.

Ameisenona, Z. 1949. Animal headed gods. *Journal of the Warburg and Courtland Institutes* 12:22–45.

Aristotle. 1965. *Historia animalum*, trans. A. L. Peck. Cambridge: Harvard University Press.

Barloy, J. J. 1974. *Man and animals*. London: Gordon and Cremonesi.

Boas, G. 1933. *The happy beast*. Baltimore: Johns Hopkins University Press.

Brion, M. 1959. *Animals in art*. London: George Harrap Co.

Carlill. 1924. *A short history of animals*. London: Kegan Paul.

Cesaresco, M. 1909. *The place of animals in human thought*. London: T. Fischer Unwin.

Celsus. 1953. Chadwick, Henry, *Origen contra Celsum*. Cambridge: Cambridge University Press.

Clark, K. 1977. *Animals and man*. New York: William Morrow.

Dante, A. 1971. *The inferno*, trans. M. Musa. Bloomington: University of Indiana Press.

Hastings, H. 1936. *Man and beast in French thought in the eighteenth century*. Baltimore: Johns Hopkins University Press.

Holt, J. 1975. *Escape from childhood*. New York: E. P. Dutton.

Homer. 1958. *The Iliad*, trans. W. Leaf and M. A. Bayfield. London: Macmillan and Co.

Klingender, F. 1971. *Animals in art and thought.* Cambridge: MIT Press.

Lovejoy, A. O., and G. Boas. 1935. *Primitivism and related ideas in antiquity.* Baltimore: Johns Hopkins University Press.

Maxwell, J. W. 1965. On action at a distance. Proceedings of the Royal Institution of Great Britain, 7. Reprinted in *Scientific papers,* ed. W. O. Niven. Cambridge 1890. Reprinted, New York: Cambridge University Press.

Mercatante, A. 1974. *Zoo of the gods.* New York: Harper and Row.

Mode Heinze. 1973. *Fabulous beasts and demons.* London: Phaidon Press.

Ruesch, H. 1978. *The social significance of animals.* Basel: Helling & Lichtenhahn.

Seton, E. T. 1901. *Lives of the hunted.* New York: Grosset and Dunlop.

Wittgenstein, L. 1958. *Philosophical investigations,* trans. G. E. M. Anscombe. 3rd ed. New York: Macmillan Co.